iWork® '09

PORTABLE GENIUS

iWork® '09

PORTABLE GENIUS

by Guy Hart-Davis

WILEY

Wiley Publishing, Inc.

iWork® '09 Portable Genius

Published by
Wiley Publishing, Inc.
10475 Crosspoint Blvd.
Indianapolis, IN 46256
www.wiley.com

Library of Congress Control Number: 2009928478

WILEY

About the Author

Guy Hart-Davis is the author of more than 50 computing books, including *iLife '09 Portable Genius* and *Mac OS X Leopard QuickSteps,* and is the coauthor of *iMac Portable Genius.*

Credits

Senior Acquisitions Editor
Jody Lefevere

Project Editor
Jama Carter

Technical Editor
Paul Sihvonen-Binder

Copy Editor
Lauren Kennedy

Editorial Director
Robyn B. Siesky

Editorial Manager
Cricket Krengel

Vice President and Group Executive Publisher
Richard Swadley

Vice President and Executive Publisher
Barry Pruett

Business Manager
Amy Knies

Senior Marketing Manager
Sandy Smith

Project Coordinator
Katie Crocker

Graphics and Production Specialists
Jennifer Henry
Andrea Hornberger
Ronald Terry

Quality Control Technician
Jessica Kramer

Proofreading
Linda Seifert

Indexing
Potomac Indexing, LLC

This book is dedicated to Rhonda and Teddy.

Acknowledgments

I'd like to thank the following people for making this book happen:

- Jody Lefevere for getting the book approved and signing me up to write it.

- Cricket Krengel and Sapna Kumar for shaping the outline.

- Jama Carter for running the editorial side of the project.

- Paul Sihvonen-Binder for reviewing the book for technical accuracy and making many helpful suggestions.

- Lauren Kennedy for copyediting the book with a light touch.

- Jennifer Henry, Andrea Hornberger, and Ronald Terry for laying out the book in the design.

- Linda Seifert for scrutinizing the pages for errors.

- Potomac Indexing, LLC, for creating the index.

Contents

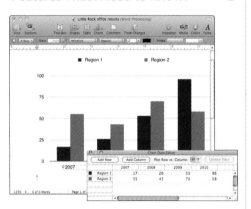

chapter 2

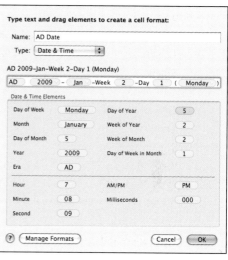

Now That I've Made My
Document, How Can I Use It? 142

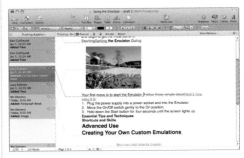

chapter 6

How Can I Work More Efficiently in Numbers? 170

chapter 7

How Do I Perform Calculations in Numbers Spreadsheets? 196

chapter 8

How Can I Make My Spreadsheets Dynamic? 220

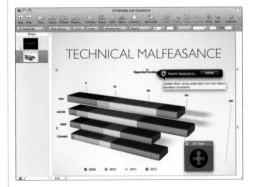

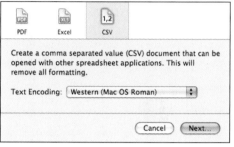

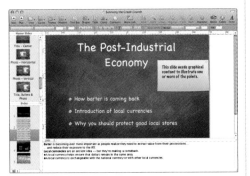

Introduction

Your Mac is a wonderful productivity tool as well as the best multimedia machine around. The iWork suite of applications lets you create slick and professional documents, spreadsheets, and presentations quickly and with minimal effort, and share them easily in person, on paper, or via the Internet.

iWork '09 Portable Genius shows you how to get the most out of the iWork applications. Here's a taste of what you can do with this book:

- **Get up to speed on all core iWork skills.** Make sure you know all the essential maneuvers you can use in any of the iWork applications — everything from customizing the toolbar and choosing shared preferences to inserting images and sounds, charts, and other objects in your documents, protecting them with passwords, and sharing them via Apple's iWork.com online service. However well you type, don't miss out on any of the time-saving ways of entering text.

- **Create professional-quality documents in Pages.** Set up the Pages window so that you can build your word-processing documents and page-layout documents fast and smoothly. Control the page layout, bring in text from outside sources, and learn how to deal with problems when exchanging documents with Microsoft Word. Format your documents quickly and consistently with styles, and make creating new documents a snap by saving your own custom templates.

- **Review and share your Pages documents.** When you need to work with colleagues to finalize a document, use the Track Changes feature to manage their editing suggestions and the Comments feature to share ideas. Finish your document by adding notes, links, and a table of contents, and use the Proofreader to catch embarrassing errors. Then share the document on the Web, by turning it into a Word document, or by creating a read-anywhere PDF file from it.

- **Build spreadsheets and crunch data in Numbers.** Pick the best template as a starting point, and then organize your data logically into sheets and tables. Use cell formatting to make Numbers format values exactly the way you want them, and save time and effort by auto-filling data, using Auto-Completion, and bringing in data from Microsoft Excel workbooks or your Address book. Make the most of Numbers' built-in functions, and create powerful formulas that perform exactly the calculations you need.

- **Make your spreadsheets powerful and persuasive.** Unvarnished data may be great for accountants, but most people prefer their data easy to read and illustrated with charts. Fully format a table in moments by applying a suitable style, or create your own custom styles to give the spreadsheets a unique and consistent look. Create charts that draw data from different sources to deliver the message you want, share them with Pages and Keynote, and even update them automatically from within a document or presentation. Add headers, footers, and page numbers to complete the spreadsheet, and then share it on the Web, as PDFs, or by exporting it to an Excel workbook or a Comma-Separated Values file.

- **Create persuasive presentations in Keynote.** Choose an effective look for your presentation by selecting the right theme — or mix and match themes to give separate parts of the presentation different flavors. Develop the presentation's outline quickly by using the Outline pane and bringing in headings from a Pages document, or simply open a PowerPoint presentation so that you can create a Keynote version of it. Use the light table to arrange your slides into the right order, set extra slides to be skipped, and add presenter notes that will help you hit every key point in order.

- **Deliver your presentation powerfully and convincingly.** Set up your presentation display to show precisely the information you need, then rehearse the presentation and make sure your timings work. Connect your Mac to an external projector or display and give the presentation, controlling it with the keyboard, the mouse, or an iPhone or iPod touch. When you can't give a presentation in person, create a presentation that runs itself — great for a kiosk or trade show — or share the presentation online using any format from Numbers to PowerPoint to PDF. You can even publish a presentation directly to YouTube from Numbers or turn it into a podcast on your iWeb site.

What Are the Common iWork Features That I Need to Know?

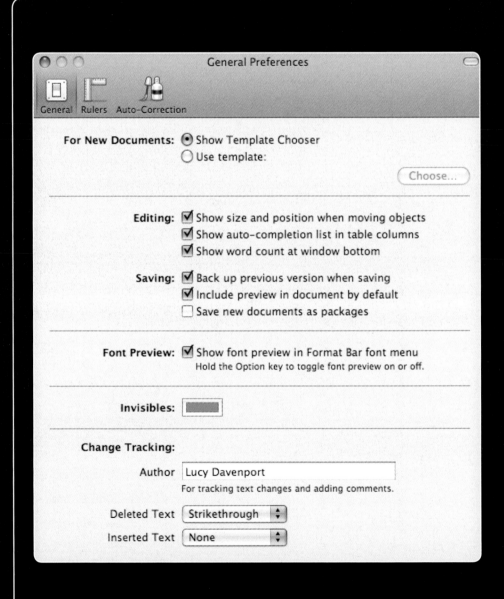

To help you create documents quickly and easily, the Pages, Keynote, and Numbers applications have many features in common. This chapter shows you not only how to customize the toolbar, add pictures and movies to your documents, choose advanced save options, and make sure your printouts match what you see onscreen but also how to share your documents using Apple's iWork .com online collaboration service and how to protect them with passwords.

Customizing the Toolbar

The toolbar in each iWork application provides a set of widely used buttons, but you'll probably want to customize it so that it contains only the buttons you need most. If your Mac has a wide screen, you can fit extra buttons on the toolbar without sacrificing any of those already there; if it does not, you can remove existing buttons to make way for others.

Which buttons you find most useful will depend on the iWork application and what you spend most time doing in it. For example, if you don't use iWork.com to share Pages documents but you do use it for Numbers and Keynote, remove the iWork.com button from Pages but keep it on the toolbar in Numbers and Keynote. Here are examples of buttons you may find useful in each application:

- **Pages.** If you often copy and paste styles, add the Copy Style and Paste Style buttons to the toolbar. If you frequently need to rearrange layers of objects, put the Front, Back, Forward, and Backward buttons on it.

- **Numbers.** If you frequently use images in your spreadsheets, add buttons such as Adjust Image, Instant Alpha, and Mask to the toolbar.

- **Keynote.** The toolbar is well populated with buttons for most purposes, but often it's useful to add the Rehearse button and Record Slideshow buttons to it so that they're immediately at hand.

Genius

If you use the iWork applications often, keep their icons in the Dock. Once you've opened an application, Ctrl+click or right-click its icon in the Dock and choose Keep in Dock. If you want Mac OS X to launch the application automatically each time you log in, Ctrl+click or right-click the Dock icon and then click Open at Login, placing a check mark next to it.

Customizing the toolbar takes only a minute:

1. **Open the application whose toolbar you want to change.** For example, click the Pages icon in the Dock to open Pages.

2. **Choose View ⇨ Customize Toolbar.** The Customize Toolbar sheet opens. Figure 1.1 shows the Customize Toolbar sheet for Pages.

3. **Click and drag each icon you want to add from the sheet to the toolbar.** Drop the icon where you want it to appear. If you want to restore the toolbar to its original set, click and drag the default set of icons from the box at the bottom of the Customize Toolbar sheet to the toolbar.

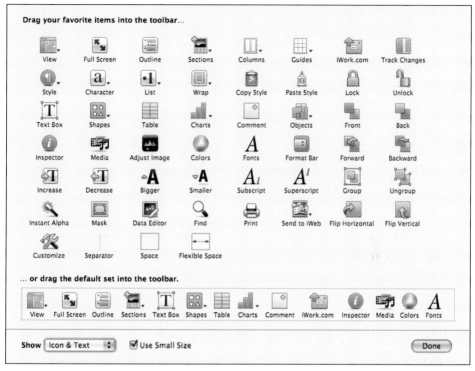

1.1 Customizing the toolbar in each iWork application takes only moments but can save you plenty of time and effort in the long run.

Genius

You can quickly customize the toolbar without opening the Customize Toolbar sheet. To reposition an item, ⌘+click and drag it to where you want it. To remove an item, ⌘+click it and drag it off the toolbar.

4. **To remove an icon from the toolbar, click and drag it off the toolbar.** You can drop it anywhere — on the Customize Toolbar sheet is often easiest — and it disappears in a puff of smoke.

5. **To move one of the existing items, click and drag it to where you want it.**

6. **In the Show pop-up menu, choose how to display the toolbar items: Icon & Text, Icon Only, or Text Only.** Icon & Text is usually clearest and easiest. Text Only gives you a shallow toolbar with more space for your documents.

7. **Select the Use Small Size check box if you want to use smaller icons and text, which lets you fit more items on the toolbar.**

8. **Click Done to close the Customize Toolbar sheet.**

Adding Your Own Templates to the Template Chooser or Theme Chooser

Pages comes with a good selection of templates for creating every kind of document from a business letter to a newsletter, or from a report to an invitation. Similarly, Numbers provides spreadsheet templates that range from tracking your weight, your workouts, and your finances to scheduling your employees, sending out invoices, and calculating return on investment. And Keynote provides an attractive variety of themes for giving presentations different looks. Even so, you'll probably want to use your own custom templates and themes as well.

You can create a template from a document in Pages or Numbers by using the File ➪ Save as Template command. Similarly, you can create a theme from a document in Keynote by choosing File ➪ Save Theme.

The iWork applications store your custom templates and themes in these folders (where the tilde, ~, represents your home folder):

- **Pages templates.** ~/Library/Application Support/iWork/Pages/Templates/Templates

- **Numbers templates.** ~/Library/Application Support/iWork/Numbers/Templates/Templates

- **Keynote themes.** ~/Library/Application Support/iWork/Keynote/Themes

If you have templates or themes from other sources, you can simply copy them to the appropriate folder using the Finder.

You can then create a new document based on a custom template or theme like this:

- **Pages and Numbers.** Choose File ➪ New from Template Chooser, and then click My Templates in the left column of the Template Chooser window.

- **Keynote themes.** Choose File ➪ New from Theme Chooser, and then scroll down to the bottom of the Theme Chooser window. Your themes appear below the built-in themes.

If you store your templates or themes in a different folder (for example, in a shared folder on your Mac or on the network), you can create a new document based on a template by choosing File ➪ Open, navigating to the template or theme, selecting it, and then clicking Open.

Choosing Preferences Common to All iWork Applications

To make the iWork applications behave the way you want them to, spend a few minutes setting preferences. For the sake of space, this section discusses preferences that all three applications have in common, or that two of the applications share, although you set them separately for each application — the settings in one application don't affect the other applications. For each application's individual preferences, see Chapter 2 for Pages, Chapter 6 for Numbers, and Chapter 10 for Keynote.

To set the preferences, open the Preferences window by pressing ⌘+, (comma) or choosing Pages ➪ Preferences, Numbers ➪ Preferences, or Keynote ➪ Preferences. Click the General button on the toolbar if the application does not display this preferences pane.

Choosing the template or theme for new documents

In the For New Documents area, choose whether to display the Template Chooser window (in Pages and Numbers) or the Theme Chooser window (in Keynote) when you start creating a new document by choosing File ➪ New or pressing ⌘+N.

If you want to use the same template or theme for each new document, select the Use template option button or the Use theme option button, click Choose, and pick the template or theme. Choose the Blank template to create a blank document in Pages or Numbers.

Choosing Editing options

For each application, select the Show size and position check box when moving objects if you want the application to display a readout showing the position of an object as you drag it. This is usually helpful for placing objects precisely. Deselecting this check box may improve performance if you're running iWork on an older Mac.

For Pages and Numbers, select the Show auto-completion list in table columns check box if you want the application to display sugges-tions for completing a cell by reusing an entry from the same column, as shown in figure 1.2.

City	Location
San Francisco	4180 Mission Street
Sacramento	1827 Lincoln Boulevard
San Jacinto	9829 Main Street
san Francisco	

San Franci...
Sacramento
San Jacinto

1.2 Auto-completion is helpful when several cells in a column contain the same entry, but it can slow down data entry when many cells have similar but not identical entries.

Keeping backups of your documents

In the Saving area of the General pane, select the Back up previous version when saving check box (or, in Keynote, the Back up previous version check box) if you want an easy way to recover your work if a document becomes corrupted. This setting is great for recovering a document from cor-ruption or an editing mistake that occurred since Mac OS X's automatic backup feature, Time Machine, last backed up the document. Beyond that, you can recover the last version of the docu-ment from Time Machine.

When this check box is selected, the application changes the name of the last version you saved to "Backup of" and the filename, then saves the current version of the document under the name. The next time you save the document, the application overwrites the last backup with the new backup.

Genius There's only ever one backup of each document, so while keeping backups does roughly double the amount of space the documents take up on your hard drive, you're not creating an endless series of separate backup files for each document.

Choosing whether to create a document preview

In the Saving area of the General pane, select the Include preview in document by default check box if you want the application to save a preview of the document in the file. The preview is a small PDF that lets Mac OS X's Quick Look feature display the document more accurately, so it helps you recognize your documents more easily. The trade-off is that the document's file size is a bit larger, but not enough to make a difference unless your Mac is flat out of disk space.

Genius

You may find that creating a preview in a document makes the Finder's Quick Look feature take longer to display the preview. This is because the preview displays a PDF with an accurate miniature picture rather than just showing a rough approximation of the document.

You can override this default setting when you're saving a document. Simply select or deselect the Include preview in document check box in the Save As dialog box.

Choosing whether to save documents as packages

Earlier versions of the iWork applications (up to and including the iWork '08 applications) saved documents in a "package" format that included all the items needed to make up the document. The package is a folder marked as a bundle, a technique widely used in Mac OS X. For example, each iWork application is a package file.

The iWork '09 applications save documents using a different format: a zipped folder that is given a different filename extension (for example, the .pages extension for a Pages document). If you want to save your iWork '09 documents as package files, select the Save new documents as packages check box in the General pane of the Preferences dialog box.

Genius

Use the package format only if you need it for compatibility with other applications. While the package format largely worked fine for the iWork '08 applications, it did create problems with some email and file transfer programs, which treated them as collections of files rather than as packages and, thus, didn't transfer documents in their entirety. If you do create a package that you need to send via email, compress it first (Ctrl+click it or right-click the file and choose Compress), and then send the resulting ZIP file.

Choosing whether to use the font preview

In the Font Preview area of the General pane of the Preferences window, select the Show font preview in Format Bar font menu check box if you want the Format bar's pop-up font menu to show each font's name in that font rather than in the system font. Using the previews lets you see that the font you're choosing is the one you want, but you see fewer fonts at once and the font pop-up menu does not appear as quickly. If you have many fonts, or your Mac is struggling to run iWork, try turning off this option.

Setting the default zoom

In Pages and Numbers, choose your preferred zoom level in the Default Zoom pop-up menu in the Rulers pane of the Preferences window. Figure 1.3 shows the Rulers pane in Pages. You can change the zoom level for any open document easily by using the Zoom pop-up menu in the status bar, but it's handy to have documents open at the zoom level you usually need.

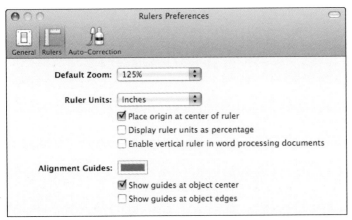

1.3 In the Rulers pane of the Preferences window, set the default zoom (for Pages and Numbers), choose your preferred ruler units and how to display them, and choose how to display alignment guides.

Choosing ruler units and ruler options

Choose the measurement units you want to use for the ruler in the Ruler Units pop-up menu in the Rulers pane of the Preferences window. For Pages and Numbers, your choices are Inches, Centimeters, Points, or Picas; for Keynote, your choices are Inches, Centimeters, or Pixels.

Note Points and picas are used in typesetting. A point is $^{1}/_{72}$ inch, and a pica is 12 points, or $^{1}/_{6}$ inch, so there are 6 picas to the inch.

Pages and Keynote provide two more shared ruler options:

- **Place origin at center of ruler.** Select this check box if you want the ruler to have 0 (zero) in the middle, with negative values to its left (−1, −2, −3) and positive values to its right (1, 2, 3). Placing the origin at the center can help when you're working on symmetrical designs for documents or slides; for asymmetrical designs, such as books and reports, the standard ruler is usually easier. If you want the ruler to appear as usual, with 0 at the left side, deselect this check box.

● **Display ruler units as percentage.** Select this check box if you want the ruler to show percentages of the available space rather than measurement units. This option can be useful when you're laying out pages or slides. You can use this setting together with the Place origin at center of ruler setting to produce a ruler that runs from 50 percent on the left through 0 percent at the midline of the document to 50 percent on the right.

Choosing an alignment guide color and object alignment options

In the Alignment Guides area of the Rulers pane, you can change the color in which the application displays alignment guides. Just click the button and use the Colors window to pick the color. Pages and Numbers use blue by default; Keynote uses blue for its master gridlines and yellow for the alignment guides.

Also in the Alignment Guides area, you can control whether the application displays the alignment guides at the center of the object, the edges, both, or neither. Experiment with different settings and find which work best for you; it'll depend on how you set up your documents and the way you prefer to work. Simply select or deselect the Show guides at object center check box and Show guides at object edges check box as needed.

Working Efficiently with Text

Whether you create your own documents or edit other people's documents, you'll almost certainly work extensively with text. This section shows you how to enter text as quickly and accurately as possible, format it using shortcuts, and personalize your custom dictionary so that it doesn't query the specialist words you need.

Time-saving ways of entering text

Regardless of how fast you type, don't ignore the other ways of entering text in a document quickly with less effort. You can use anything from iWork's built-in features to optical character recognition software or dictation software.

Making the most of Auto-Correction's substitution

To enter text quickly and accurately, you'll want to exploit the Auto-Correction substitution feature in each of the iWork applications. Auto-Correction monitors the characters you type, and when you type a predefined entry and then press the spacebar or another key that indicates the end of a word, Auto-Correction replaces what you've typed with the correct text. Auto-Correction also replaces straight quotes with smart (curly) quotes, among several other tricks.

Genius

The name *Auto-Correction* suggests the feature is only for correcting mistakes, but you can save much more time by creating Auto-Correction entries that replace your custom abbreviations with full words. For example, if you write about management, creating an entry named "mgt" that expands to "management" saves you 7 keystrokes, while "tmgt" for "time management" saves you 11 keystrokes.

You'll need to turn Auto-Correction's substitution feature on in each application, and then build a list of entries that contain the text you want. Each application has its own list of entries, so you can use the same abbreviation to insert different text in different applications.

Each application comes set with a handful of corrections for common typos (for example, it substitutes "the" when you type "teh") and symbols (for example, it substitutes the ½ symbol when you type 1/2 and the © symbol when you type (c)). You can add further entries as needed. An entry can be either a single word or multiple words. For example, you could create an entry to change "should of done" to "should have done."

Here's how to turn Auto-Correction's substitution feature on:

1. **Open the application's Preferences window.** Press ⌘+, (comma) or open the application's menu and choose Preferences. For example, choose Pages ➪ Preferences.

2. **Click the Auto-Correction button to display the Auto-Correction pane.** Figure 1.4 shows the Auto-Correction pane for Keynote. Pages and Numbers have all the same options except for Underline text hyperlinks on creation.

3. **Select the Symbol and text substitution check box.**

Note

While you have the Auto-Correction pane open, make sure the other settings — for smart quotes, automatic capitalization, superscripting suffixes such as 1st and 2nd, automatically detecting email addresses and Web addresses, and automatically detecting lists — are set to suit the way you work.

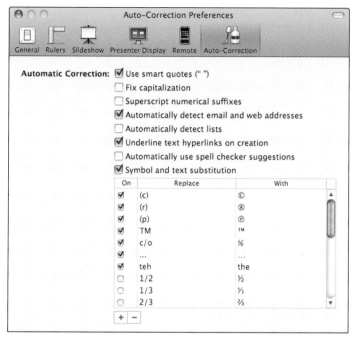

1.4 Turn on symbol and text substitution in the Auto-Correction pane of the Preferences window.

Keep the Preferences window open so that you can add entries like this:

1. **Click the Add (+) button at the bottom of the pane.** The Preferences window adds a new row at the bottom of the list and selects its check box.

2. **Type the misspelling or abbreviation in the box in the Replace column, then press Tab to move to the box in the With column.** Make sure that an abbreviation isn't a real word — unless you actually want to replace that word each time you type it.

3. **Type or paste the correct spelling in the box in the With column, and then press Return.** The Replace text can be several words or sentences, but Auto-Correction preserves only its text. Any formatting or other items (such as a paragraph mark in Pages or the division between cells in Numbers) are lost.

4. **Repeat Steps 1 through 3 to add as many Auto-Correction entries as you need, and then close the Preferences window.**

If you want to stop using an Auto-Correction entry for a while, simply deselect the check box alongside it. To get rid of an entry, click it, and then click the Delete (-) button at the bottom of the pane.

Genius

Unlike applications such as Microsoft Word, the iWork applications don't let you quickly create Auto-Correction terms from mistakes you uncover when checking your spelling. But it's well worth taking the extra time to open the Preferences window and add an entry for any spelling error you may make again. You can save effort by copying the mistake and pasting it into the Preferences window.

Entering text with automatic typing utilities

Auto-Correction is great for when you're working in the iWork applications, but if you want to use the same abbreviations to enter text automatically in any application on your Mac, you need to look elsewhere. Two utilities that let you enter text via abbreviations are TypeIt4Me (www .typeit4me.com) and TextExpander (www.smileonmymac.com/TextExpander/).

Pasting text without formatting

If you have an electronic version of the text you want to use, you can simply paste it into your iWork document.

When you paste normally by using the Paste command on the Edit menu or the shortcut menu, the pasted material retains the original formatting. To remove this formatting and have the pasted material assume the formatting of the paragraph, cell, or placeholder in which you paste it, choose the Paste and Match Style command on the Edit menu or the shortcut menu instead.

Genius

The easiest way to execute the Paste and Match Style command is to press ⌘+Option+Shift+V.

Entering text by inserting a file

Some applications let you insert an entire document inside another document. The iWork applications don't offer this feature, but you can insert all the contents of one document into another document by opening the source document, selecting and copying its contents, and then pasting it into the destination document.

Entering text via optical character recognition

If you need to enter hard copy text from a printed document into an iWork document, you have three main choices: you can type the text, scan the document and use optical character recognition (OCR) software, or dictate the text.

If you have a scanner that includes OCR software, you're ready to go. Otherwise, you'll need to add scanning software. Scanning software is typically expensive, so it may not be worth buying unless you have many pages to process. At the time of this writing, the leading OCR applications for the Mac are Readiris Pro for Mac (www.irislink.com) and OmniPage Pro X for Macintosh (www.nuance.com/omnipage/mac/).

Once you've scanned the text, proofread it carefully and clean up any errors. You can then paste the text into your iWork document.

Entering text using dictation software

If your typing is slow, or if you need to enter huge amounts of text, dictation software can save you time and effort. Like OCR software, dictation software is expensive; but unlike OCR software, you can use it for creating any document by wearing a headset microphone and speaking aloud.

At this writing, the leading dictation software for Mac OS X is MacSpeech Dictate (www.macspeech.com), which uses the speech-recognition engine from the widely respected Dragon NaturallySpeaking application. Depending on the dictation software you choose, you can either dictate into the application's text editor (and then paste the text into the iWork application) or dictate directly into the iWork application.

Genius

Dictation software never spells a word incorrectly, but it often substitutes the wrong word or phrase for what you've said. This means it's a good idea to review your work shortly after you dictate it so that you can straighten out any substitutions. Otherwise, you may be pretty sure that you didn't say a particular phrase but find it hard to remember what you actually said.

Keyboard shortcuts for formatting text

You can format text using the Format bar or the Font panel, but you'll often be able to work more quickly and efficiently using keyboard shortcuts. Table 1.1 lists the keyboard shortcuts you should know for all the iWork applications.

Table 1.1 iWork Keyboard Shortcuts for Formatting Text

Formatting	Keyboard Shortcut
Toggle boldface	⌘+B
Toggle italics	⌘+I
Show or hide the Fonts window	⌘+T
Increase the font size	⌘++

continued

15

Table 1.1 continued

Formatting	Keyboard Shortcut
Decrease the font size	⌘+−
Apply or remove superscript	⌘+Control++
Apply or remove subscript	⌘+Control+−
Align text left	⌘+{
Align text right	⌘+}
Center text	⌘+|
Justify text	⌘+Option+|

Personalizing your custom dictionary

Like most Apple applications, the iWork applications use Mac OS X's main dictionary files for checking spelling and let you add the extra words you need to your custom dictionary file. This file has the name "en" if you're using U.S. English and is located in the ~/Library/Spelling folder (where ~ represents your home folder).

Adding words to the custom dictionary

The typical way to add words to the custom dictionary is by adding them one by one while checking spelling:

- If you're using the Check Spelling as You Type feature, Ctrl+click or right-click a word that the spelling checker has queried and choose Learn Spelling from the shortcut menu.
- If you've opened the Spelling window, click the Learn button.

You can also add words by using the application described in the next section.

Removing words from the custom dictionary

Checking the spelling in a document that contains many specialized terms can become so mind-numbing that it's easy to add a word to the dictionary by mistake. To avoid the spelling checker happily accepting this word from now on, you need to remove the word from the custom dictionary.

Removing a word is harder than it should be, because while the dictionary file is a text file that you can open with an application such as TextEdit, it uses a special character to mark the end of each word. TextEdit can't display this character, so you can't be sure you've deleted it.

The best solution is to download the free Dictionary Cleaner application from Two AM Software (http://twoamsoftware.com/?q=dc/about). Dictionary Cleaner installs as a preference pane in System Preferences and provides an easy-to-use interface for adding new terms, correcting existing mistakes, or removing words you no longer want in the dictionary.

Genius

Instead of using Dictionary Cleaner, open your dictionary file in a more powerful text editor than TextEdit — one that can display the hidden delimiter character (for example, BBEdit from Bare Bones Software). It's worth experimenting if you need to add a long list of terms to the dictionary. After saving and closing the dictionary, you need to log out and then log back in to make Mac OS X notice that you've changed the dictionary.

Giving Your Documents Punch with Photos and Images

A great way to give your documents punch is to insert photos or images in them. The iWork applications let you quickly add items from your Mac's media library or from its file system.

Inserting a photo or image

Inserting a photo from your media library takes only a moment:

1. **Open the Media Browser.** Click the Media button on the toolbar or choose View ➪ Show Media Browser.

2. **Click the Photos tab (see figure 1.5) if either of the other tabs is displayed.**

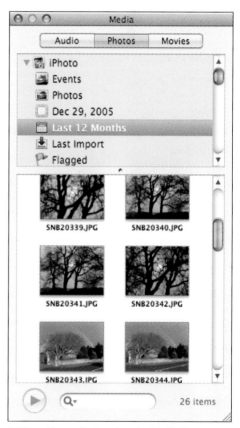

1.5 Inserting a photo using the Media Browser.

3. **Find the photo you want.** For example, click the album or event that contains it, or use the search box at the bottom of the Media Browser to search for it by name or keyword.

4. **Click the photo, drag it to your document, and drop it where you want it to appear.**

If you've got iPhoto open, you don't need to use the Media Browser. Just click the photo you want, drag it to the iWork application's window, and drop it where you want it to appear.

You can also add a picture from a Finder window by clicking and dragging. Again, just drop the picture where you want it to appear in the iWork document.

Resizing and masking a photo or image

To make the photo or image the size you need, you can resize it by selecting it and then clicking and dragging one of the handles around it. Drag any handle — corner or side — to resize the image proportionally.

If you don't want the whole of the photo or image to appear, you can mask it. *Masking* means hiding part of the image so that it's not visible in the document. Masking is similar to cropping, except that you don't crop off the hidden part of the photo: it's still there, which means you can reveal parts of it as needed.

Caution

Masking is usually helpful, but when you share a document with other people, make sure the masked parts of the images don't contain anything you don't want those people to see. When you share the document, someone can remove the masking and see the hidden parts of the image. If the masked parts of your images contain sensitive data, create cropped versions of the images and insert those in your document instead.

Here's how to mask an image:

1. **Click the image, and then choose Format ⇨ Mask or click the Mask button on the toolbar (in Keynote).** The application displays a rectangular mask over the middle of the image, together with the masking controls (see figure 1.6).

2. **Resize the mask as needed by dragging its sizing handles.** Shift+click and drag to resize the mask proportionally.

3. **If necessary, rotate the mask by ⌘+clicking and dragging a sizing handle.** Similarly, you can rotate the image by ⌘+clicking and dragging one of its sizing handles.

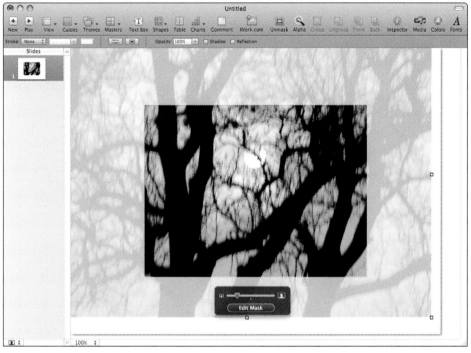

1.6 Mask an image to hide the parts you don't want to see in your document.

4. **To change which part of the image appears through the mask, move the mouse pointer inside the mask so that it appears as a hand.** Click and drag the image to change the part shown in the mask. Click and drag the size slider in the masking controls if you want to zoom the image in or out.

5. **To hide the masked part of the image so that you can see the effect you've produced, click Edit Mask.** Click again if you want to display the mask once more.

6. **When you've adjusted the mask and the image to your satisfaction, click elsewhere in the document to deselect the image.**

Adjusting the photo or image to make it look the way you want

If you need to adjust the colors of the photo or image to make it suit your document better, choose View ⇨ Show Adjust Image to open the Adjust Image window (see figure 1.7), and then use the controls in it to change the color balance. Table 1.2 lists the tools in the Adjust Image window and what they do.

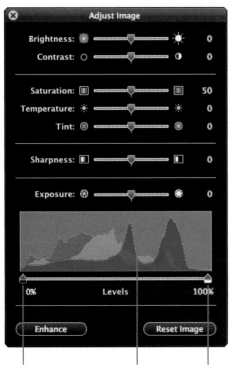

Black slider   Levels histogram   White slider

1.7 The Adjust Image window lets you quickly change the color balance of a photo to suit your document.

Masking an Image with a Shape

A regular rectangular mask works well for many images, but sometimes you may prefer to use another shape — for example, a circle or a triangle.

To mask an image with a shape, click the image, choose Format⇨Mask with Shape, and then choose the shape from the submenu. Your choices are Rectangle, Rounded Rectangle, Oval, Triangle, Right Triangle, Arrow, Double Arrow, Diamond, Quote Bubble, Callout, Star, or Polygon.

Once you've applied the shape mask on the image, you can adjust the mask and the image under it as needed.

If you need to remove the mask, click the image and choose Format⇨Unmask.

Table 1.2 Tools in the Adjust Image Window

Tool	Effect
Brightness slider	Increases or decreases the overall brightness of the image.
Contrast slider	Adjusts the contrast of the image.
Saturation slider	Changes the intensity of the color.
Temperature slider	Changes the color temperature. The left end of the slider gives a "cool" blue effect, and the right end gives a "warm" golden effect.
Tint slider	Changes the amount of green and red tones in the image. Drag to the left to add red and reduce green; drag to the right to reduce the red and add green.
Sharpness slider	Increases the sharpness of the image, making it look crisper.
Exposure slider	Adjusts the brightness of the image.
Levels histogram	Shows how the colors in your picture are distributed between pure black (at the left end, 0 percent) and pure white (at the right end, 100 percent). The red, green, and blue show the individual red, green, and blue color channels in the image.
Black slider	Adds black tones to the image (drag the slider to the right).
White slider	Adds white tones to the image (drag the slider to the left).
Enhance button	Attempts to make the image look better by analyzing its colors and adjusting them automatically. Enhancing an image often improves it; if not, you can reset the image.
Reset Image button	Resets all the sliders to how they were when you started editing.

Removing the background from a photo or image

The Instant Alpha tool in the iWork applications lets you make parts of an image transparent. You can use this to remove from an image any parts that you don't want to see in your document, which creates a very dramatic effect.

To fade out parts of the background, follow these steps:

1. **Click the image, and then choose Format ⇨ Instant Alpha from the menu bar.** In Keynote, you can also click the Alpha button on the toolbar.

2. **Zoom in if necessary so that you can see clearly what you're doing.**

3. **Click in the color of the background you want to remove, and then drag to increase the size of the circle that the application displays.** As you drag, the application fades out the color you chose and colors in the same range. Figure 1.8 shows an example.

Center of circle   Selected color region
Edge of circle   Crosshair

1.8 To remove a background, turn on the Instant Alpha tool, and then click and drag with the crosshair. The application gradually selects the color range as the circle grows.

4. **Click and drag elsewhere if you need to remove further sections of color or further colors.**

5. **When you've finished removing the colors, press Return to close the Instant Alpha tool.**

Note

If you need to remove an alpha effect, click the image, and then choose Format ⇨ Remove Instant Alpha.

Rotating a photo or image

You can quickly rotate a photo, image, or other object. Click the object to select it, then:

- To rotate the object freely, ⌘+click one of the selection handles, and then drag it to the angle you want. The application displays a readout of the angle you've reached.

Genius

If a photo from your photo library needs rotation, it's usually easier to rotate the photo in iPhoto before bringing it into an iWork document.

- To rotate the object in 45-degree stages, hold down Shift as well as ⌘, then click and drag.

- To rotate the object freely about a handle rather than about its center, ⌘+Option+click and drag the opposite handle. For example, to rotate the object pivoting on the upper-left corner, ⌘+Option+click the lower-right handle and drag.

- To rotate the object in 45-degree stages about a handle, ⌘+Option+Shift+click and drag the opposite handle.

Putting a frame on a photo or image

If you want to make a photo or image stand out from its surroundings, you can add a frame to it like this:

1. **Click the photo or image to select it.**

2. **Click the Inspector button on the toolbar to open the Inspector window, then click the Graphic Inspector button.**

3. **Open the Stroke pop-up menu, then choose Picture Frame.**

4. **Use the other controls in the Stroke area (see figure 1.9) to choose the type of frame, color, scale, or size.** The options available depend on the type of frame you choose.

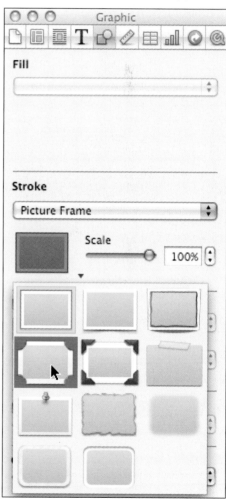

1.9 You can quickly add a custom frame to a photo or image using the Graphic Inspector.

Adding preset shapes and creating custom shapes

The iWork applications provide a useful variety of preset shapes — lines, arrows, squares, triangles, and so on. To insert a shape in a document, you simply click the Shapes pop-up menu on the toolbar and click the shape you want. You can also insert shapes from the Insert ➪ Shape submenu.

Once you've placed a shape, you can resize it, rotate it, change its color, or add a frame to it by using the techniques discussed earlier in this chapter.

To create a custom shape, click the free-form icon at the bottom of the Shapes pop-up menu or choose Insert ➪ Shapes ➪ Draw a Shape. Draw the shape you need by clicking to place each control point (see figure 1.10).

Press Esc when you've finished drawing the shape. You can then adjust the shape by clicking and dragging one of the red control points.

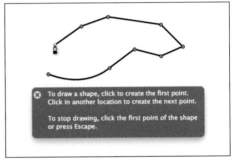

1.10 You can draw your own custom shapes using the Draw a Shape tool.

Inserting text in a shape

If a shape has space inside it, as circles do, you can insert text in the shape. Simply double-click inside the shape to place the insertion point in it, and then type or paste in the text. You can then edit and format the text as normal.

Inserting Movies and Sounds

You can insert movies and sounds in your iWork documents by using much the same techniques as for photos:

- **Open the Media Browser and click the Audio tab or the Movies tab.** Click the sound file or movie file you want, and then drag it to your document.

- **Open a Finder window to the folder that contains the sound file or movie file.** Click the file, and then drag it to your document.

- **Choose Insert ➪ Choose from the menu bar (or press ⌘+Shift+V) to open the Choose dialog box.** Click the file, and then click Insert.

◉ **If the document contains a media placeholder, click and drag a file to it from the Media Browser or a Finder window.** Using a media placeholder has two main differences from placing an object freely: first, the position for the file is set already (although you can change it if you need to); and second, if you click and drag a second file, it replaces the file already in the placeholder.

Once you've inserted a movie or sound in your document, choose QuickTime settings for it to control how it plays. Follow these steps:

1. **Click the movie or sound object in the document to select it.**

2. **Click the Inspector button on the toolbar to open the Inspector window.** Click the QuickTime Inspector button to display the QuickTime Inspector (see figure 1.11).

3. **Click and drag the Start slider to set the frame at which to start playing the movie or the point at which to start playing the sound.** For a movie, the document displays the current frame as you drag. If you want to play the whole movie or sound so that you can see or hear it, use the playback controls at the bottom of the QuickTime Inspector.

4. **Click and drag the Stop slider to set the frame at which to stop playing the movie or the point at which to stop playing the sound.**

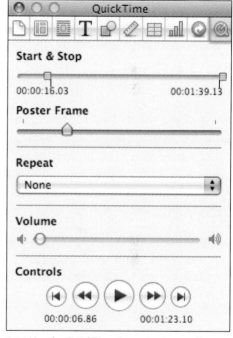

1.11 Use the QuickTime Inspector to specify which part of the movie or sound file to play, the playback volume, and whether to loop it.

5. **For a movie, click and drag the Poster Frame slider to pick the frame you want to display in the document until the movie starts playing.** You may want to use the opening title as the poster frame, but for other movies, you may find another frame is more colorful, dramatic, and compelling.

6. **In the Repeat pop-up menu, choose whether to loop the movie.** Choose None to play it once, Loop to play it continuously until the viewer stops it, or Loop Back and Forth to play it forward, then backward, and so on.

7. **Click and drag the Volume slider to set the playback volume.** Full volume may give someone opening the document an unpleasant surprise.

Adding Charts

A chart is a great way of adding detail, impact, and persuasion to a document. While Numbers is the main charting application in the iWork suite, you can easily add charts to your Pages documents and Keynote presentations as well.

Choosing the right chart type for your data

To make your chart convey your interpretation of the data most clearly, you need to choose the right chart type. Table 1.3 lists the main chart types in detail and when each type is effective. For all chart types except the Scatter chart, iWork also offers a 3D version that works the same way but adds greater visual impact (sometimes at the expense of clarity).

Table 1.3 iWork's Main Chart Types

Chart Type	Explanation	When to Use It
Column	Each value appears as a separate vertical column.	To show how the values compare to each other.
Stacked Column	Each category appears as a separate vertical column, divided into separate colored sections that represent each value.	To show the contribution each item makes in a category. For example, to show how different products contribute to your company's revenues.
Bar	Each value appears as a separate horizontal bar.	To show how the values compare to each other, using a horizontal orientation rather than a vertical orientation.
Stacked Bar	Each category appears as a separate horizontal bar, divided into separate colored sections that represent each value.	To show the contribution each item makes in a category.
Line	Each value appears as a point, with a line connecting the points and representing the category as a whole.	To show changes over time — for example, to chart temperature or fluctuations in a currency's value.

Chart Type	Explanation	When to Use It
Area	Each category appears as a line connecting the points (values), but the area below the line is shaded.	To show the contribution of each series over time.
Stacked Area	Like an area chart, but with the areas stacked on top of each other to make them more visible.	As with the area chart, to show the contribution of each series over time, but to provide extra visual clarity.
Pie	Each value in a single series of data appears as a slice of a round pie.	To show how each value contributes to the whole (the pie).
Scatter	Each value appears as a separate point.	To show how two different sets of numbers are related, either without drawing a line through the data points or by drawing a "best-fit" line. Often used for representing medical or scientific studies.
Mixed	The chart presents two data series together on a single chart. For example, you can combine a line chart and a column chart.	When you need to show two separate sets of data that can use the same axes.
2-axis	The chart represents each of its two data series as a separate chart. The two charts share an X axis but have separate Y axes, one on the left and one on the right.	To show two separate sets of data that require different Y axes but can share an X axis (for example, showing years or departments).

Inserting a chart

You insert a chart in Pages and Keynote using a different technique than in Numbers. In Pages and Keynote, you use the Chart Data Editor, a mini spreadsheet window, to enter and edit the chart data, while in Numbers you enter the chart data in a table in a spreadsheet.

Inserting a chart in Pages or Keynote

To insert a chart in Pages or Keynote, click the Charts pop-up menu on the toolbar and click the chart type you want. Keynote adds the chart to the document and displays the Chart Data Editor (shown in figure 1.12) and the Chart Inspector.

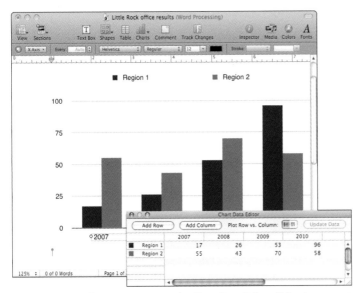

1.12 In Pages and Keynote, you use the Chart Data Editor dialog box to set up the data for a chart.

You can now add your data to the chart like this:

- **Replace the sample data.** Double-click a column label or row label, and then type the label you want to use. Similarly, double-click a sample value, and then type your real value in its place.

- **Paste in data from Numbers or Excel.** In Numbers or Excel, select the cells containing the data, and then copy them (for example, press ⌘+C). In the Chart Data Editor, click in the upper-right cell, and then press ⌘+V to paste in the data.

- **Add a row or column.** Click Add Row or Add Column. Replace the row or column label and the default values with your data.

- **Delete a row or column.** Click the row heading or column heading, and then press Delete.

- **Move a row or column.** Click the row heading or column heading, and then drag it to where you want it to appear. The Chart Data Editor moves the other rows or columns to accommodate the one you dragged.

As you work in the Chart Data Editor, the application updates the chart in your document.

Inserting a chart in Numbers

To insert a chart in Numbers, enter the chart data in a table, select the table, click the Charts pop-up menu on the toolbar, and then click the chart type you want. You can also choose Insert ⇨ Chart and then click the chart type on the Chart submenu.

To change the chart's data, edit the table on which you based the chart.

Changing a chart's orientation

If you realize that you've laid out your data the wrong way — you've entered the data series in rows when you should have entered it in columns, or vice versa — you can quickly fix the problem:

- **Pages or Keynote.** Click the Plot Rows as Series button or the Plot Columns as Series button (found on the left and right, respectively, next to Plot Rows vs. Column) in the Chart Data Editor.
- **Numbers.** Click the table, and then click the button that appears at its upper-left corner.

The iWork application then automatically switches the series in the chart to show the columns or rows. You don't need to retype the data as you do in many other applications.

Formatting a chart

To format a chart, use the controls in the Chart Inspector (see figure 1.13). Normally, the iWork applications open the Chart Inspector when you insert a chart, so it may be open already. If not, click the Inspector button on the toolbar, and then click the Chart Inspector button in the Inspector window.

Resizing and repositioning the chart

To resize the chart, click it, and then drag one of the selection handles. To reposition the chart, click it and drag to where you want it to appear.

For more precise placement, use the Metrics Inspector, as discussed later in this chapter.

1.13 The Chart Inspector is your one-stop shop for formatting a chart.

Changing the chart colors

You can change the color for a single chart series or for all the chart series at once. Follow these steps:

1. **In the Chart Inspector, click Chart Colors to open the Chart Colors window (see figure 1.14).**

2. **In the upper pop-up menu, choose the type of fill you want — for example, 3D Texture Fills.**

3. **In the lower pop-up menu, choose the category of fills — for example, Bright Color or Etched Metal.**

4. **Change the series colors as needed.** You can then either click a fill from the box and drag it to the chart series to which you want to apply it, or click Apply All to apply as many of the fills needed for all the series in your chart.

5. **Click the Close button (the red button) to close the Chart Colors window.**

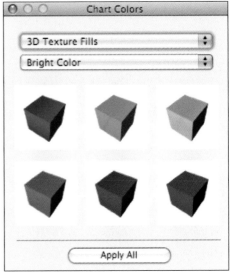

1.14 The Chart Colors window lets you quickly change the colors used in the chart.

Displaying axes and borders

Use the options in the Axis pane in the Chart Inspector to control which axes and borders appear. The options depend on the chart type you're using, but for most types, you can choose which axes to display and whether to display borders around the chart.

Choosing number formats

The formatting controls in the Axis pane in the Chart Inspector let you choose how numbers appear. The selection of controls changes to reflect the choice you make in the Format pop-up menu, but these are the controls you'll use most of the time:

● **Format.** In this pop-up menu, choose the number format: Number, Currency, Percentage, Date and Time, Duration, Fraction, or Scientific. If you need to create a custom number format, click Custom, and then work on the sheet that appears.

○ **Symbol.** If the format needs a symbol before it, open this pop-up menu and choose the symbol. For example, for a Currency format, you can choose a symbol such as US Dollar or Euro. If the format doesn't need a symbol, the Symbol pop-up menu doesn't appear.

○ **Decimals.** If the number format includes decimal places, you can adjust the number of decimal places in this text box. The box doesn't appear for formats that don't use decimal places, such as fractions.

○ **Separator.** If the number format requires a separator character (such as a decimal point), make sure this check box is selected.

○ **Negative format.** In this pop-up menu, choose between using a minus sign to denote a negative number and using parentheses around the number.

○ **Suffix.** If the number format requires a character placed after the number, type it in this box.

Formatting 3D settings

When you create a 3D chart, the Chart Inspector adds the 3D Scene pane (see figure 1.15) at the bottom to let you control the 3D effects. You can do the following:

○ Click a direction arrow and drag to change the 3D positioning of the chart. You can also drag a direction arrow in the little 3D Chart pop-up window that the application displays.

○ Open the Lighting Style pop-up menu and choose the lighting style you want — Default, Glossy, Medium Center, Medium Left, Medium Right, Soft Fill, or Soft Light.

○ Click and drag the Chart Depth slider to make the chart the depth you want.

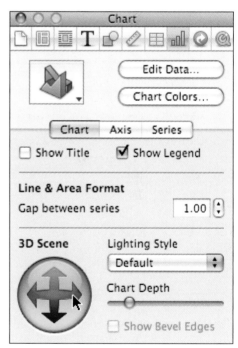

1.15 You can adjust the 3D positioning, lighting style, and chart depth for a 3D chart.

Adding a legend and labels

To make a chart easy to read, you can add a *legend*, a key to the different colors the chart uses. To display the legend, select the Show Legend check box in the Chart pane of the Chart Inspector. You can then click the legend and drag it to another position if you want. To format the legend, click it, and then use the controls on the Format bar or in the Text Inspector as for any other text.

iWork adds labels to most charts automatically, but you'll often need to adjust them to show exactly what you want. To change the labels displayed for the axes, use the controls in the Axis pane. Figure 1.16 shows the available options in the Value Axis (X) pop-up menu.

Changing a chart from one chart type to another

If you find you've chosen the wrong chart type to convey the meaning you want your data to give, you can quickly change to another chart type. Simply open the Chart Inspector window, open the Chart Type pop-up menu, and then click the chart type you want.

1.16 Choose options for axis labels, tick marks, and gridlines in the Axis pane of the Chart Inspector.

After the application applies the new chart type, you may need to resize the chart or change options to suit the new type.

Working with Objects

Object is the catch-all term for images, movies, sounds, shapes, and other separate items that you place in your documents. You've seen many of the essential techniques for working with objects earlier in this chapter.

This section makes sure you know all the essential maneuvers for objects — selecting them, rotating and flipping them, copying and moving them, arranging them in stacks, and more. This section mostly uses images as examples because they tend to be the most widely useful objects.

Selecting an object

You can select objects in iWork documents by using the techniques you know from working in Mac OS X:

- Click the first object to select it.
- Add further objects to the selection by Shift+clicking or ⌘+clicking them.
- Add a range of objects by Shift+dragging or ⌘+dragging around them.
- ⌘+click a selected object to remove it from the selection.

Rotating and flipping objects

You can rotate any object by one of its handles using the following click-and-drag options:

- ⌘+click and drag to rotate the object freely.
- ⌘+Shift+click and drag to rotate the object in 45-degree stages.
- ⌘+Option+click and drag to rotate the object freely about the handle opposite the one you click.
- ⌘+Option+Shift+click and drag to rotate the object in 45-degree stages about the handle opposite the one you click.

To flip an object horizontally, click it and choose Arrange ➪ Flip Horizontally. To flip an object vertically, click it and choose Arrange ➪ Flip Vertically. Repeat the flipping if you want to restore the object to its previous orientation.

Genius

You can also rotate and flip objects using the Angle and Flip controls in the Rotate section of the Metrics Inspector.

Copying and moving objects

You can copy an object by Option+clicking it and dragging to where you want the copy to appear. Alternatively, you can use the Copy and Paste commands as usual. For example, Ctrl+click or right-click an object, and then choose Copy from the shortcut menu to copy it; Ctrl+click or right-click the destination, and then choose Paste from the shortcut menu.

You can move an object any of these ways:

- **Mouse.** Click the object and drag it.

- **Arrow Keys.** Select the object and Down Arrow, Up Arrow, Right Arrow, or Left Arrow to move it by one point. To move by ten points, hold down Shift and press the appropriate arrow key.

- **Metrics Inspector.** Click the object, display the Metrics Inspector, and then change the values in the X box and Y box in the Position area.

Arranging overlapping objects to show what you want

When you place one object on top of another, each object obscures the objects underneath it. To make objects appear the way you want them, you shuffle them into the right order.

To move an object forward or backward in the stack of objects, Ctrl+click or right-click the object and then choose Bring Forward, Bring to Front, Send Backward, or Send to Back from the shortcut menu. You can also use the following shortcuts:

- **Move to the back.** Press ⌘+Shift+B

- **Move back one layer.** Press ⌘+Option+Shift+B

- **Move to the front.** Press ⌘+Shift+F

- **Move forward one layer.** Press ⌘+Option+Shift+F

Aligning and spacing objects

When you're placing two or more objects, you can use the alignment guides (see figure 1.17) to position objects relative to each other. But iWork also includes automated features to help you align objects accurately and distribute them evenly.

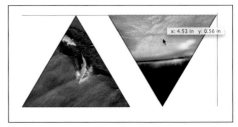

1.17 iWork's alignment guides help you align objects accurately.

To align selected objects, Ctrl+click or right-click in the selection, highlight Align Objects on the shortcut menu, and then click the alignment you want: Left, Center, or Right for horizontal alignment; or Top, Middle, or Bottom for vertical alignment.

To space out selected objects evenly, Ctrl+click or right-click in the selection, highlight Distribute Objects on the shortcut menu, and then choose Horizontally or Vertically, as appropriate.

Placing an object exactly with the Metrics Inspector

To adjust the placement of a selected object precisely, open the Metrics Inspector by clicking the Inspector button and the toolbar and then clicking Metrics Inspector. The Metrics Inspector's controls (see figure 1.18) are largely self-explanatory. Select the Constrain proportions check box if you want the object to retain its original proportions as you change its size.

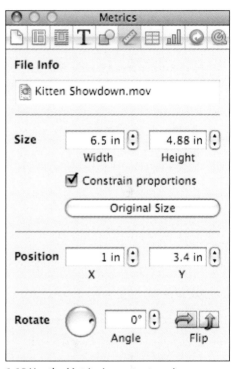

1.18 Use the Metrics Inspector to resize, reposition, or rotate an object precisely.

Adding connection lines

When you've placed two objects in the same document, you'll often need to connect them with a line. You can use a line shape easily enough, but the iWork applications offer a better option: you can use a custom connection line, which iWork associates with the objects and repositions automatically if you move the objects.

To connect two objects with a connection line, follow these steps:

1. **Select the objects.** Click the first object, and then ⌘+click the other object.
2. **Choose Insert ➪ Connection Line.** The application inserts a straight connection line from one object to the other.

35

3. **If you want to bend the connection line, as shown in figure 1.19, click the white editing point in the line's middle and drag.**

4. **If you want to move an end of the line away from the shape it touches, click the blue circle and drag it to where you want.**

5. **Use the controls on the Format bar to change the formatting of the line as needed.** For example, you can change the line weight or color, or add arrows to the ends of the line.

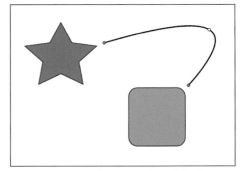

1.19 You can quickly insert and modify a line connecting two objects. Drag the white circle to bend the line, or drag the blue circles to move the ends away from the edges of the objects.

Adding shadows and reflections

From the Graphic Inspector, you can add a shadow to most objects. Click the object, select the Shadow check box, and then use the controls in the Shadow area to set the color, angle, offset, blur, and opacity of the shadow.

You can add a reflection to the bottom of graphical objects (such as pictures). Select the Reflection check box in the Graphic Inspector to apply the reflection, and then click and drag the slider to set the strength of the reflection.

Adding a fill, gradient, or picture to a shape

You can enhance objects such as text boxes and shapes by adding a fill, gradient, or picture.

Click the object, display the Graphic Inspector, and then choose one of the following from the Fill pop-up menu:

- **None.** Choose this item to remove the existing fill.

- **Color Fill.** Click the Fill Color button to open the Colors window, choose the color, and then click the Close button (the red button).

- **Gradient Fill.** Click the Start Gradient Fill Color button to open the Colors window and choose the starting color for the gradient fill (see figure 1.20). Click the End Gradient Fill Color button, choose the ending color for the gradient fill, and then click the Close button. Use the Angle controls to set the angle for the fill: you can drag the round knob, click the down-pointing arrow to set the gradient angle to 270 degrees, click the right-pointing arrow to set the gradient angle to 0 degrees, or type the number of degrees in the box.

Start Gradient Fill Color button

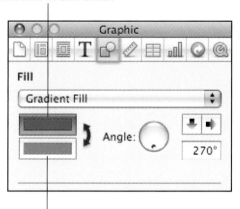

End Gradient Fill Color button

1.20 Setting up a color gradient fill in the Graphic Inspector.

- **Advanced Gradient Fill.** Use the Start Gradient Fill Color button, Change Gradient Fill Color button, and End Gradient Fill Color button at the bottom of the slider (see figure 1.21) to open the Colors window and choose the colors for the gradient fill. Click the slider to place further Change Gradient Fill Color buttons, and choose colors for them too. Drag the Gradient Blend Point markers to adjust the points at which blending starts. Choose between a linear gradient and a radial gradient by clicking the Linear Gradient button or the Radial Gradient button. If you need to reverse the gradient, click the Reverse the Gradient's Direction button. Use the Angle controls to set the angle for the fill.

- **Image Fill.** Click the image file in the Open dialog box that appears, and then click Open. In the Image Size pop-up menu, choose how to size the image: Scale to Fit, Scale to Fill, Stretch, Original Size, or Tile. Scale to Fill is usually the best choice, as it displays as much of the image as possible, as large as possible, without leaving white space.

- **Tinted Image Fill.** Choose the image file and image size as described previously. You can then click the Colors button to open the Colors window and select the colors with which to tint the image.

Gradient Blend Point marker   Change Gradient Fill Color button

Start Gradient Fill Color button   End Gradient Fill Color button

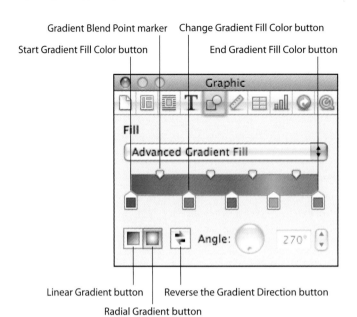

Linear Gradient button   Reverse the Gradient Direction button

Radial Gradient button

1.21 Setting up an advanced gradient fill in the Graphic Inspector.

Locking and grouping objects

The iWork applications make objects so easy to move that you sometimes need to prevent yourself from moving them.

To keep an object in place, you can lock it. Simply click the object, and then choose Arrange ➪ Lock or press ⌘+L. If you need to unlock a locked object, click it, and then choose Arrange ➪ Unlock or press ⌘+Option+L.

When you need to keep several objects in the same positions relative to each other, you can group them. Select the objects, and then choose Arrange ➪ Group or press ⌘+Option+G. The application then deselects the individual objects and puts a single selection box around the group. You can then move the group as you would a single object (or, if you prefer, lock it in place).

To ungroup grouped objects, click the group, and then choose Arrange ➪ Ungroup or press ⌘+Option+Shift+G. The application removes the single selection box and selects the individual objects.

Making an object partly transparent

To make an object partly transparent, drag the Opacity slider in the Graphic Inspector. 100% opacity is fully opaque; 0% opacity is fully transparent. As you drag the slider to change the opacity, you'll see objects under the object you're working on become more visible or less so.

Choosing Advanced Save Options

You save a document in iWork by choosing File ⇨ Save as usual, but the iWork applications also offer two advanced Save options that you can use when you want to share the document with another Mac. These options let you make sure the document contains the movie files, audio files, and template components needed to appear in all its glory on another Mac.

Here's how to use the advanced options:

1. **Expand the Save As dialog box if necessary.** If the Save As dialog box is in its small state, not showing the folders and options, click the disclosure triangle to the right of the Save As field to expand the dialog box and reveal them.

2. **Click the Advanced Options disclosure triangle to display the Advanced Options section of the dialog box (see figure 1.22).**

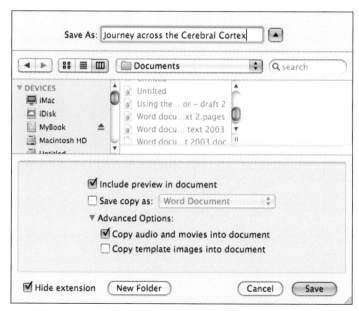

1.22 When saving a document to use on another Mac, use the Advanced Options section in the Save As dialog box to make sure the document contains all the component files it needs.

3. **Add the audio and movies.** Make sure the Copy audio and movies into document check box is selected when you want to save the audio and movies in the document rather than playing them from their files on your Mac. Saving the files in the document makes the document bigger — usually adding the size of the audio or movie files, plus a little overhead — but ensures that the files will play on another Mac. If you don't save the files in the document, they'll play only if the other Mac has the files in the same places in its file system — possible if you synchronize two of your Macs, but unlikely with anyone else's Mac.

4. **Add the template images.** Select the Copy template images into document check box if you want to include in the document any images used in the template. Including the images makes sure they appear on any other Mac that opens the document. Normally, you need to select this check box only when you're using one of your own templates rather than one of the standard iWork templates, unless whoever's using the other Mac has removed the standard templates from it.

Note The Copy template images into document option is selected but unavailable when you're saving a document as a template in Pages or Numbers. This is because the template must contain its own images.

5. **Finish saving the document as usual.** Type the filename, choose the folder, and then click Save.

Printing Your Documents

As with most Mac applications, you can print from the iWork applications by choosing File ⇨ Print or pressing ⌘+P, choosing options in the Print dialog box, and then clicking Print.

You can pick the printer you want from the Printer pop-up menu, choose the number of copies and whether to collate them, and decide whether to print all the document's pages or just a range of pages.

Beyond these basic options, you will probably want to exploit Mac OS X's printing options for layout, paper handling, cover pages, and so on. And if you print color documents, you should use ColorSync profiles to make your printouts match the colors you see on-screen.

Note Beyond these common options, each iWork application has its own set of printing options for its document type. For example, Keynote lets you choose whether to print slides, slides with notes, an outline, or a handout. I'll show you the most important options in the chapters on the individual applications.

Exploiting the most useful print options

You can access the following print options from the pop-up menu in the middle of the Print dialog box:

- **Layout.** These options let you print multiple pages of the document per sheet of paper instead of a single page. When printing four or more pages per sheet, you can choose among various layout directions to control the order in which the pages appear on the sheet.

- **Paper Handling.** Instead of printing all the pages in order, you can print just the odd pages or just the even pages. This can be useful when you need to duplex copies. You can scale the printout to fit the page size, and print the pages in reverse order if needed. You can print a page border (useful for separating multiple pages on the same sheet), turn on two-sided printing if your printer supports it, and reverse the page orientation.

- **Cover Page.** This option lets you choose whether to print a cover page to separate your print job from others. You can print the page before or after the document, and you can include billing information to help you track costs.

- **Scheduler.** You can schedule jobs for printing at particular times — for example, when your colleagues aren't using the printer — or put them on hold. You can also assign a priority (Urgent, High, Medium, or Low) to a print job for when you're competing with your colleagues.

- **Paper Type/Quality.** Choose the paper type (for example, plain, letterhead, or labels) and the tray from which to pull it on the printer. Depending on the printer, you may also be able to print in a draft mode to save ink.

Using ColorSync to make your printouts match your on-screen colors

To make the most of the images and colors that the iWork applications let you include in your documents, you'll probably want to make sure the colors you get in your printouts are the same as the colors you see on-screen. This doesn't happen automatically, but you can use ColorSync profiles to make the colors match.

Changing to a different ColorSync profile

Here's how to change to a different ColorSync profile:

1. **Choose Apple menu ⇨ System Preferences to open System Preferences.**

2. **In the Hardware section, click Displays to open the Displays pane.**

3. **Click the Color tab to display its contents (see figure 1.23).** The Display Profile box shows the list of profiles. Select the Show profiles for this display only check box if you want to view only the profiles designed for the display you're using.

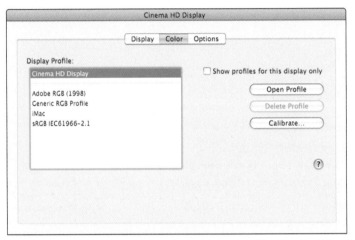

1.23 You can quickly change to another ColorSync profile using the Color tab in the Displays pane in System Preferences.

4. **Click the profile you want to apply.** You'll see the colors on screen change, either sub-tly or more dramatically.

5. **If you find a profile that suits you, leave it selected and quit System Preferences.** Otherwise, switch back to your original profile, keep System Preferences open, and cre-ate a custom color profile, as described next.

Creating a ColorSync profile

If none of the ColorSync profiles suits you, create a custom one that does:

1. **From the Color tab of the Displays pane in System Preferences, click Calibrate.** The Display Calibrator Assistant opens (see figure 1.24).

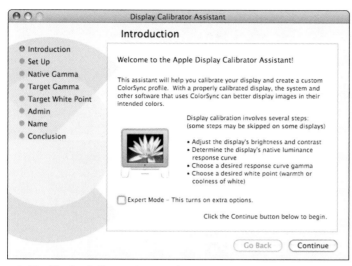

1.24 The Display Calibrator Assistant lets you create a custom ColorSync profile to ensure that colors appear the way you want them.

2. **Select the Expert Mode check box if you want to set the full range of options.** Deselect this check box if you want to skip the steps for determining the display's native luminance response curves.

3. **Click Continue, and then follow through the screens, choosing the settings that look best to you.** You can click Go Back at any point to return to a previous setting and change it.

4. **At the end of the process, give your new color profile a descriptive name so that you can easily identify it, and then click Done to close the Display Calibrator Assistant.** Mac OS X automatically applies your new profile as you create it.

5. **Choose System Preferences ⇨ Quit System Preferences if you've finished working with color profiles.**

Managing colors when printing

Here's how to make your printouts match your preferred color profile:

1. **When your document is ready for printing, choose File ⇨ Print (or press ⌘+P) to open the Print dialog box as usual.**

2. **If the Print dialog box appears at its small size, click the disclosure triangle to the right of the Printer pop-up menu to expand the dialog box to its full size.**

3. **Choose Color Matching in the pop-up menu in the middle of the dialog box.** The Color Matching options appear (see figure 1.25).

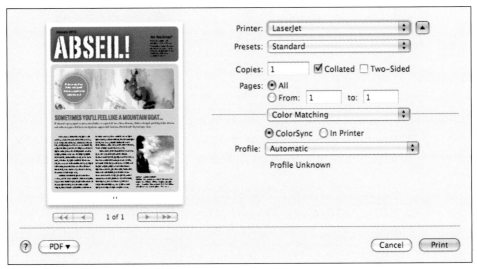

1.25 Use the Color Matching options in the Print dialog box to make sure your printouts use the right color profile.

4. **Select the ColorSync option button rather than the In Printer option button.**

5. **Choose Other Profiles from the Profile pop-up menu to open the Select ColorSync Profile dialog box (see figure 1.26).**

6. **Click the profile you want to use, and then click OK to close the Select ColorSync Profile dialog box.** The profile appears in the Profile pop-up menu.

7. **Choose other printing options as needed, and then click Print to print the document.**

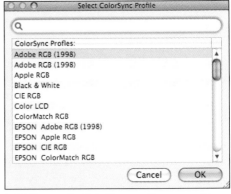

1.26 Choose the ColorSync profile you want to use for printing the document.

Genius

To keep the iWork applications running as well as possible, install any updates as soon as Apple releases them. To check for updates, choose Apple menu ⇨ Software Update, which will offer you any available patches for Mac OS X and the Apple software you're running. If you don't check via Software Update, the iWork applications will prompt you when an update is available.

Sharing Documents Using iWork.com

If you want to be able to share your iWork documents easily online with other Mac users, see if Apple's iWork.com online collaboration service meets your needs. Apple has integrated iWork .com right into the iWork applications, so it could hardly be easier to use.

Genius

To use iWork.com, you must have an Apple Account. If you have a MobileMe subscription or you've ever bought from the iTunes Store or the App Store, you've already got an account. If not, you can easily create one.

Here's how to share a document via iWork.com:

1. **Create the document as usual in Pages, Numbers, or Keynote, and then save it on your Mac or a network drive.** If you want to protect the document with a password, do so as described in the next section.

2. **With the document still open, click the iWork.com button on the toolbar.** The application displays the Sign In to share documents via iWork.com dialog box (see figure 1.27).

1.27 You must sign in to iWork.com in order to share documents. If you don't already have an Apple Account, you can create one.

3. **Sign in, or create an account and then sign in.**

○ If you don't have an Apple Account, click Create New Account and follow through the process of setting up an account.

○ Once you're armed with an Apple Account, enter your Apple ID and password in the boxes, and then click Sign In.

4. **The application displays the Invite others to view a copy of your document via iWork.com dialog box.** Figure 1.28 shows this dialog box with its Advanced section displayed and options chosen.

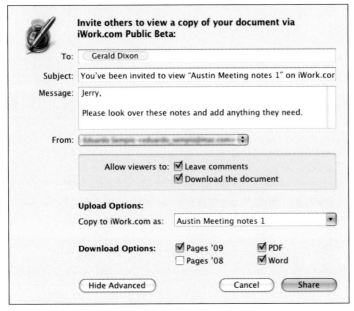

1.28 Sharing a document via iWork.com

5. **Type the address in the To box. Mac OS X automatically suggests addresses from your Address Book as you type.**

6. **Type a subject in the Subject box and any message needed in the Message box.**

7. **Make sure the From pop-up menu is showing the email address from which you want to send the message.** Choose a different address if necessary.

8. **In the Allow viewers to area, select the Leave comments check box, the Download the document check box, or both, to control what the people with whom you're sharing the document can do to it.**

9. **Click Show Advanced to display the Upload Options and Download Options areas of the dialog box.**

10. **In the Copy to iWork.com as box, you can change the filename under which the document is shared.** For example, you may want to enter a more explicit name for the shared document than the name on your Mac.

11. **In the Download Options area, choose which formats to provide for download (if you selected the Download the document check box).** You can choose among Pages '09, Pages '08, PDF, and Word.

12. **Click Share.** The application displays the Copying document to iWork.com dialog box as it uploads the document, and then displays the Your document can now be viewed on iWork.com dialog box (see figure 1.29).

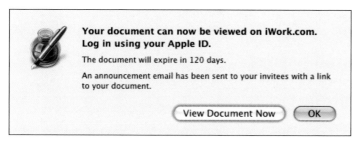

Your document can now be viewed on iWork.com. Log in using your Apple ID.

The document will expire in 120 days.

An announcement email has been sent to your invitees with a link to your document.

View Document Now   OK

1.29 After sharing a document on iWork,com, it's a good idea to check it briefly to make sure it appears as you want it to.

13. **Click View Document Now if you want to check the document, or click OK if you just want to dismiss the dialog box.**

iWork.com also sends you an email message containing a link to the shared document. You can click this link to open the document in your Web browser and see what comments viewers have left.

Similarly, when someone else invites you to view a document on iWork.com, you receive an email message. When you click the link to view the document, and then log into iWork.com, you can view the document online and (if the sharer has allowed you to) leave comments or download the document.

Protecting a Document with a Password

Before you share an iWork document, you may want to protect it with a password to keep inquisitive eyes off your sensitive data. Here's how to apply a password:

1. **With the document open, display the Document Inspector.** Click the Inspector button on the toolbar, and then click Document Inspector.

2. **Select the Require password to open check box at the bottom.** The application displays the Set a password to open this document dialog box (see figure 1.30).

3. **Type the password in the Password box and the Verify box.** If you want help choosing a memorable or hard-to-crack password, click the key icon to the right of the Password box to launch the Password Assistant (see figure 1.31). Choose the password type — for example, Memorable or Random — in the Type pop-up menu, then drag the Length slider to pick the number of characters you want. The Suggestion box shows a suggested password, and the Quality bar shows roughly how hard the password is to crack.

1.30 Password protecting an iWork document gives you a modest level of protection against casual snooping.

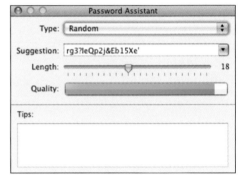

1.31 The Password Assistant can help you devise truly fiendish passwords to protect your documents.

4. **If you must, type a hint in the Password Hint box.** Having a password hint compromises your password's security, so it's far better not to create one.

5. **Click Set Password.**

6. **Save the document.** For example, press ⌘+S.

To remove a password, deselect the Require password to open check box at the bottom of the Document Inspector. You'll need to enter the password before the application lets you remove the protection. Once you've done so, save the document.

Caution Password-cracking utilities are widely available on the Internet and can strip away most password protection in seconds. iWork's protection is fine for normal use, but you shouldn't rely on it to keep your data safe against a determined attacker.

How Can I Work Faster in Pages?

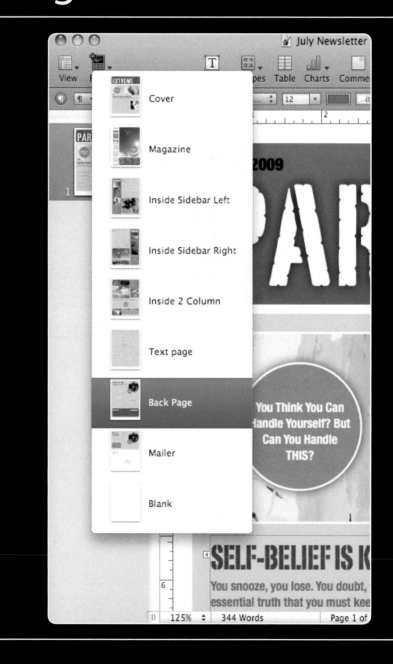

Pages makes it easy for you to create powerful and persuasive documents. To create those documents quickly and efficiently, you need to understand Pages' components and what they do, set preferences to suit the way you work, and customize the Pages window to show the tools you need. This chapter also covers what you need to know about controlling page layout, adding text and new template pages to a document, and bringing Microsoft Word documents into Pages.

Knowing What You Are Working With

When you open Pages and choose the document type in the Template Chooser window (if it appears), you should see something like figure 2.1.

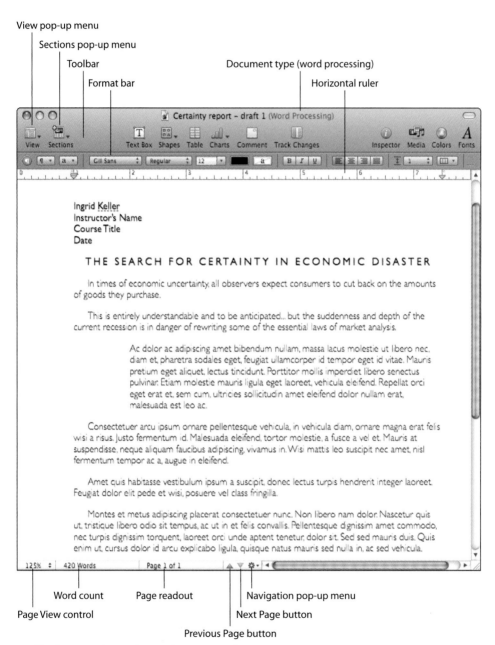

2.1 The Pages window with a word-processing document open.

Here's the lowdown on what these elements do:

- **Document type.** Shows whether the document is a word-processing document or a page-layout document. More on this in a moment.

- **Toolbar.** Provides buttons and pop-up menus for common actions and for displaying other tools (such as the Inspector window). You can customize the toolbar as discussed in Chapter 1.

- **Format bar.** Gives you access to the most widely used formatting commands (such as styles, font formatting, and alignment) and to the Styles drawer.

- **Horizontal ruler.** Lets you control the horizontal placement of text, tabs, and other elements.

- **View pop-up menu.** Lets you change the view and the Pages elements displayed. For example, you can display or hide the rulers.

- **Sections pop-up menu.** Lets you quickly add another section to the document. The sections available depend on the template on which the document is based.

- **Page View control.** Lets you zoom in and out, and change the number of pages displayed in the Pages window.

- **Page readout.** Shows the current page number and the total number of pages in the document.

- **Previous and Next buttons.** Let you move quickly from page to page or to the other item you've chosen in the Navigation pop-up menu.

- **Navigation pop-up menu.** Lets you choose how to navigate your document: by pages or sections; by moving to a footnote or endnote, to a bookmark, to a hyperlink, to a merge field, or to a comment; or by moving to an instance of a paragraph style or a character style.

Genius

When you're working on a word-processing document, the vertical ruler appears only if you've selected the Enable vertical ruler in word processing documents check box in the General pane of the Preferences window for Pages.

As you see in the Template Chooser window, Pages divides documents into two categories:

- **Word-processing documents.** These are documents — such as reports, letters, and resumes — in which most of the text is in one main section that flows from page to page as needed.

- **Page-layout documents.** These are documents — such as newsletters, brochures, posters, or business cards — that have text in various discrete sections, mostly in separate text boxes. Page-layout documents also tend to contain more pictures and other graphical elements than word-processing documents, but there are plenty of exceptions to this rule.

When you have a page-layout document open, as in figure 2.2, the Pages window includes the following items:

- **Vertical ruler.** Helps you position items accurately on the page.

- **Thumbnail view.** This sidebar shows a thumbnail of each page. You click the thumbnail to move quickly to the page.

- **Pages pop-up menu.** Lets you quickly add pages from the page-layout template to the document. The pages available depend on the template — for example, a newsletter template may provide a back page and a mailer as well as pages with different numbers of columns. This pop-up menu replaces the Sections pop-up menu on the toolbar.

Pages pop-up menu   Document type (page layout)

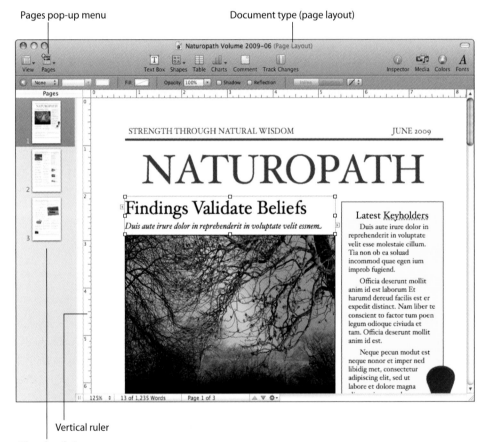

Vertical ruler

Thumbnail view

2.2 With a page-layout document open, the Pages window automatically displays the vertical ruler and the thumbnail view.

Setting Pages-Specific Preferences

Apart from the common preferences discussed in Chapter 1, Pages lets you choose settings for hidden characters, for displaying the ruler, and for tracking changes to text. To set the preferences, open the Preferences window by pressing ⌘+, (comma) or choosing Pages ⇨ Preferences. If the General pane isn't displayed, click the General button to display it.

Choosing an Invisibles color

In the Invisibles area, choose the color in which you want Pages to display its *invisibles*. These are formatting marks — such as spaces, tabs, and paragraph breaks — that normally are invisible.

Choosing ruler settings

Select the Enable vertical ruler in word processing documents check box if you want to display the vertical ruler in word-processing documents as well as page-layout documents. Normally, you won't need the vertical ruler in word-processing documents.

Setting the author name

Make sure the Author box is showing the name you want to assign as the author of the document, comments, and tracked changes. If you're creating documents for yourself, this will presumably be your own name or a pseudonym, but if you're creating documents for someone else, you may need to use their name instead.

Choosing Track Text Changes settings

Pages' Track Changes feature lets you set Pages to mark the insertions and deletions in the document. Chapter 5 shows you how to use Track Changes.

In the Deleted Text Style pop-up menu, choose how deleted text should appear. Your choices are Underline, Strikethrough, and None. Strikethrough is usually the most helpful choice, but choosing None lets you see how the document will read when the text has been deleted — it's still there, but has no marking. (You can also turn off the display of tracked changes temporarily, which has a similar effect.)

In the Inserted Text Style pop-up menu, choose the formatting you want Pages to apply to new text inserted in the document. Again, your choices are Underline, Strikethrough, and None. Underline is usually the most helpful choice, as Strikethrough is confusing. (It makes text appear to be deleted). You may prefer to use None when you're editing the document and want to see the current version of text without seeing the details of the changes themselves.

Customizing the Pages Window for Faster Work

Apple sets up Pages so that it's easy to work with, but you'll almost certainly be able to save time and effort by customizing the window to suit the documents you create, the tools you use, and the way you work.

Your first step is to customize the toolbar as discussed in Chapter 1. After that, adjust the zoom, choose a working mode, and pick the best mix of screen elements, as described here.

Zooming in and out

Use the Page View control in the lower-left corner of the Pages window (see figure 2.3) to zoom in or out so that you see as much of the document as you need to. Choose between One Up (viewing one page) and Two Up (viewing two pages) at the top of the menu, and then choose a zoom level. Your options are from 25% (tiny) up to 400% (huge).

You can also zoom the view by using the View ⇨ Zoom submenu, which provides Zoom In, Zoom Out, Actual Size, Fit Width, and Fit Page commands. From the keyboard, press ⌘+> to zoom in to the next preset zoom level or ⌘+< to zoom out to the next preset level.

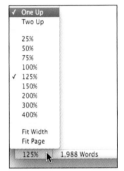

2.3 Choose the number of pages (One Up or Two Up) and the zoom level from the View pop-up menu.

Genius

The Fit Width zoom level is often the best choice for working on word-processing documents, while Fit Page is useful for working on page layouts. If you find you need to zoom each document you open in or out, set your preferred zoom percentage in the Default Zoom pop-up menu on the Rulers pane in Pages' Preferences window (Pages ⇨ Preferences). This menu doesn't include the Fit Width and Fit Page choices.

Switching among Pages' views

Pages provides four views of your documents:

- **Writing view.** This view shows your document's contents but not the outlines of the text areas.

- **Layout view.** This view shows the outlines of your document's text areas, so you can easily see the headers, footers, text boxes, and other elements.

- **Outline view.** This view shows your document as a collapsible outline of headings. You can choose how many levels of headings to display. See Chapter 3 for details of Outline view.

- **Full Screen.** This view shows your document full screen, so that you can concentrate on it without distractions.

As the names suggest, writing view is best for writing and editing documents, while layout view is good for adjusting their layout. Outline view is designed for developing the structure of a document, and full screen view is intended to let you work with fewer distractions.

You can switch between writing view and layout view in any of these ways:

- **Toolbar.** Open the View pop-up menu and choose Show Layout or Hide Layout.

- **Keyboard.** Press ⌘+Shift+L.

- **Menu bar.** Choose View ⇨ Show Layout or View ⇨ Hide Layout.

To enter full screen view, click the Full Screen button on the toolbar, choose View ⇨ Enter Full Screen, or press ⌘+Option+U. To exit Full Screen view, move the mouse up to the top of the screen and either click the Exit button on the bar that appears or choose View ⇨ Exit Full Screen. Alternatively, press ⌘+Option+U again.

To switch to outline view, click the Outline button on the toolbar or choose View ⇨ Show Document Outline. To switch back, click the Outline button again or choose View ⇨ Hide Document Outline.

Choosing which screen elements to display

Pages lets you display or hide various parts of its window to suit the document you're working on and the changes you're making to it. Figure 2.4 illustrates most of these parts of the Pages window — but not all of them, because you can't display them all at once.

Styles drawer | Thumbnail view | Toolbar | Comments pane | Format bar

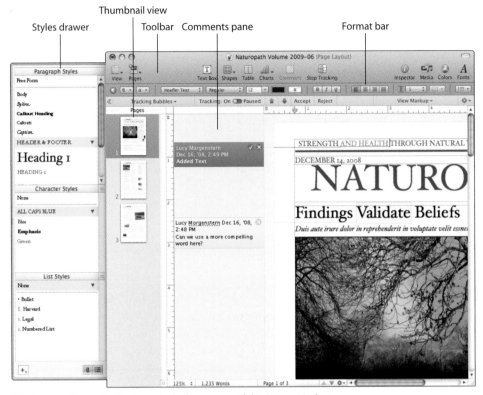

2.4 You can choose to display or hide these parts of the Pages window.

The following sections explain the parts of the window and how to display and hide them.

Toolbar

The toolbar across the top of the Pages window provides buttons for frequent actions, such as choosing which screen elements to display, inserting text boxes and tables, and opening the Inspector window. You can customize the toolbar with your favorite buttons as discussed in Chapter 1.

You can also hide the toolbar if you need more screen space. To display or hide the toolbar, choose one of these ways:

- **Mouse.** Click the jellybean-shaped button at the right end of the title bar.
- **Keyboard.** Press ⌘+Option+T.
- **Menu bar.** Choose View ⇨ Hide Toolbar or View ⇨ Show Toolbar.

Format Bar

The Format bar appears under the toolbar (when the toolbar is displayed) and provides quick access to the Styles drawer and widely used formatting commands.

You can display or hide the Format bar in these ways:

- **Mouse.** Click the View pop-up menu on the toolbar and choose Show Format Bar or Hide Format Bar.
- **Keyboard.** Press ⌘+Shift+R.
- **Menu bar.** Choose View ⇨ Hide Toolbar or View ⇨ Show Toolbar.

Rulers

Pages automatically displays both the horizontal ruler and vertical ruler in page-layout documents. In word-processing documents, Pages displays the horizontal ruler in layout view.

You can also display and hide the rulers manually in these ways:

- **Mouse.** Click the View pop-up menu on the toolbar and choose Show Rulers or Hide Rulers.
- **Keyboard.** Press ⌘+R.
- **Menu bar.** Choose View ⇨ Show Rulers or View ⇨ Hide Rulers.

Styles drawer

This slide-out drawer lets you apply, create, and manage styles (see Chapter 3 for details). You can show or hide the Styles drawer in these ways:

- **Mouse.** Click the blue Styles Drawer button at the left end of the Format bar. Alternatively, click the View pop-up menu on the toolbar and choose Show Styles Drawer or Hide Styles Drawer.
- **Keyboard.** Press ⌘+Shift+T.
- **Menu bar.** Choose View ⇨ Show Styles Drawer or View ⇨ Hide Styles Drawer.

Search sidebar

The search sidebar lets you search for all instances of a word or phrase in your document, and view a list of the results. You can click a result in the list to view it on the page.

You can display or hide the search sidebar in these ways:

- **Mouse.** Click the View pop-up menu on the toolbar and choose Search.
- **Keyboard.** Press ⌘+Option+F.
- **Menu bar.** Choose View ⇨ Search.

Comments pane

The Comments pane appears at the left side of the window and shows the comments and the tracked changes in the document. Pages automatically displays the Comments pane when you turn on Track Changes and when you insert a comment.

You can display or hide the Comments pane in these ways:

- **Mouse.** Click the View pop-up menu on the toolbar and choose Show Comments or Hide Comments.
- **Menu bar.** Choose View ⇨ Show Comments or View ⇨ Hide Comments.

Thumbnail view

Thumbnail view displays a sidebar showing a thumbnail picture of each page to help you navigate through your document. Click a thumbnail to go to its page.

You can display or hide the thumbnail view in these ways:

- **Mouse.** Click the View pop-up menu on the toolbar and choose Page Thumbnails.
- **Keyboard.** Press ⌘+Option+P.
- **Menu bar.** Choose View ⇨ Show Page Thumbnails or View ⇨ Hide Page Thumbnails.

Note

Thumbnail view and the search sidebar occupy the same space on the screen, so you can display only one of them at once. Displaying the search sidebar closes thumbnail view if it is open, and vice versa.

Invisible characters

Normally, Pages doesn't display any characters for the following formatting marks, which it calls *invisibles*:

- **Spaces.** These include spaces, nonbreaking spaces, and tabs. A nonbreaking space is one you put between words (by pressing Option+Space) to prevent Pages from breaking a line between them. Nonbreaking spaces are useful for names and technical terms, but you can use them on any words you don't want broken across lines.

Breaks. These include line breaks, paragraph breaks, page breaks, column breaks, layout breaks, and section breaks.

Anchor points. An anchor point indicates where an inline object (such as a shape or image) is anchored.

You can toggle the display of invisibles in these ways:

Keyboard. Press ⌘+Shift+I.

Menu bar. Choose View ➪ Show Invisibles or View ➪ Hide Invisibles.

Toolbar. Open the View pop-up menu and choose Show Invisibles or Hide Invisibles, as appropriate.

Controlling the Page Layout

Each of Pages' templates includes a full page layout, so you may simply be able to choose the template you need and use it without making any further adjustments. But if you need to create a custom document, you may need to change the page size and orientation, adjust the margins, add headers and footers, or insert page numbers.

Setting the page size and orientation

The easiest way to set the page size, orientation, and margins of a document is to work from the Document Inspector. Click the Inspector button on the toolbar to display the Inspector window, and then click the Document Inspector button to display the Document Inspector (see figure 2.5). If the Document pane isn't displayed, click the Document tab.

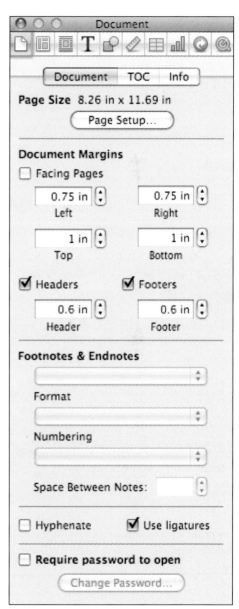

2.5 The Document Inspector gives you one-stop control over your document.

From here, you can change the page size and orientation of the document like this:

1. **Click the Page Setup button to open the Page Setup dialog box (see figure 2.6).**

Note If you're not starting from the Document Inspector, choose File ⇨ Page Setup to open the Page Setup dialog box.

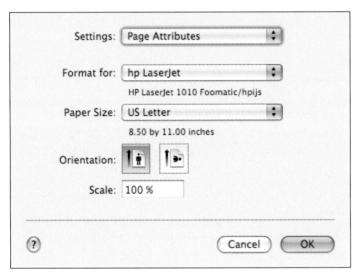

2.6 Use the Page Setup dialog box to choose the printer and set the paper size and orientation for the document.

2. **In the Format for pop-up menu, choose the printer you'll use for the document.** Choose a printer even if you're not planning to print the document.

3. **In the Paper Size pop-up menu, choose the paper size for the document.** For example, choose US Letter if you're using paper that's 8.5 × 11 inches.

Genius If the Paper Size pop-up menu doesn't include the paper size you want, click Manage Custom Sizes at the bottom of the menu to open the Custom Page Sizes dialog box. Click the + button to create a new size called Untitled, type a name for it, and then set the page size and the margins. Click OK when you've finished. Pages automatically selects the new paper size in the Paper Size pop-up menu.

4. **In the Orientation area, click the appropriate orientation button.** Click the Portrait orientation button (the left button) for paper longer than it is wide, or the Landscape orientation button (the right button) for paper that's wider than it is long.

5. **If you want to print the document larger or smaller than its actual size, increase or decrease the value in the Scale box.** This value is 100% — regular size — by default.

6. **Click OK to close the Page Setup dialog box.**

Leave the Document Inspector open so that you can set the page margins, as described next.

Setting the page margins

To set the page margins for your document, use the controls in the Document Margins area of the Document pane in the Document Inspector:

1. **If your document is designed with facing pages, select the Facing Pages check box.** When you do this, Pages changes the Left box to an Inside box and the Right box to an Outside box.

2. **Set the values for the side margins.** If your design doesn't have facing pages, change the values in the Left box and Right box. If your design does have facing pages, change the values in the Inside box and the Outside box.

3. **Set the values for the top and bottom margins.** Adjust the values in the Top box and the Bottom box.

4. **If your document will have a header, select the Headers check box, and then adjust the value in the Header box as needed.**

5. **If your document will have a footer, select the Footers check box, and then adjust the value in the Footer box as needed.**

6. **Close the Document Inspector.** Either click its red Close button or click the Inspector button on the Pages toolbar.

Adding headers and footers

Most documents benefit from having headers and footers — text that appears at the top of each page (the header) or at the bottom (the footer).

Adding simple headers and footers

To add a header to the document, select the Headers check box on the Document pane of the Document Inspector (as described in the previous section). You can then click in the header area at the top of the page (see figure 2.7) and type the header. You can also open the View pop-up menu and choose Show Layout to reveal the header and footer areas.

In the same way, you can add a footer by clicking in the Footers area.

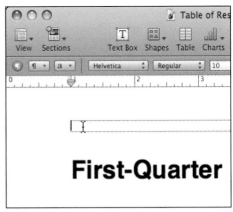

First-Quarter

2.7 Pages displays a border around the header area when you move the mouse pointer over it.

Creating different headers or footers on different pages

Once you've added a header or footer to a document, Pages automatically continues the header or footer on each page in the document. This is often convenient, but in many documents you'll need to have different headers and footers on different pages.

If you want to put a different header or footer on the first page or on the left and right pages of the document, follow these steps:

1. **Open the Layout Inspector, and then display the Section pane (see figure 2.8).** Click the Inspector button on the toolbar to open the Inspector window, click the Layout Inspector button, and then click the Section tab.

2. **In the Configuration area, select the First page is different check box if you want a different header and footer on the first page.**

3. **Also in the Configuration area, select the Left and right pages are different check box if you want different headers and footers on the left pages than on the right pages.**

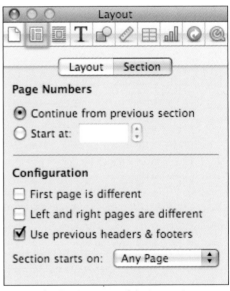

2.8 Use the Section pane of the Layout Inspector to set up different headers and footers on the first page and on left and right pages.

If you want to have a different header and footer, or sets of different left-page and right-page headers and footers in different parts of your documents, you need to break the document into sections, as described later in this chapter.

You can then choose either to continue the previous section's headers and footers or create new ones. To continue them, click in the section, and then select the Use previous headers & footers check box in the Configuration area of the Section pane in the Layout Inspector. To create new headers and footers, click in the section, deselect this check box, and then enter the headers and footers you want.

Adding page numbers

Pages makes it easy to add page numbers to your headers and footers. You can either use the Auto Page Numbers feature or insert page numbers manually.

Here's how to add page numbers using Auto Page Numbers:

1. **Choose Insert ⇨ Auto Page Numbers.** Pages displays the Insert Page Numbers dialog box (see figure 2.9).

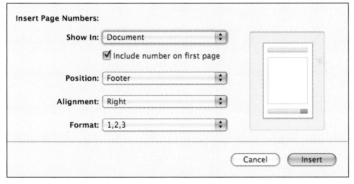

2.9 The Insert Page Numbers dialog box lets you quickly add standard page numbers to a document or a section. The blue rectangle shows where the page numbers will appear.

2. **In the Show In pop-up menu, choose where to put the page numbers.** Choose Document to add them throughout the document. Choose Current Section to add them only to the current section.

3. **Deselect the Include number on first page check box if you don't want a page number on the first page of the document or section.**

4. **In the Position pop-up menu, choose Header or Footer, as appropriate.**

5. **In the Alignment pop-up menu, choose the alignment: Left, Center, Right, Inside, or Outside.** Inside and Outside are for facing-page designs.

6. **In the Format pop-up menu, choose the numbering type you want: 1, 2, 3; a, b, c; A, B, C; i, ii, iii; or I, II, III.**

7. **Click Insert. Pages closes the dialog box and inserts the numbers.**

Genius

You can use the Auto Page Numbers feature to move your existing page numbers to a different place or change their format. Simply choose Insert ⇨ Auto Page Numbers and use the Insert Page Numbers dialog box as before.

If you prefer to insert page numbers manually, simply click where you want the page number to appear, and then choose Insert ⇨ Page Number.

What you'll often want is to create Page X of Y numbering — for example, Page 4 of 20 — that lets the reader see where he is in the document. To do so, type **Page** and a space, choose Insert ⇨ Page Number, type **of** with spaces before and after, and then choose Insert ⇨ Page Count.

Genius

To change the formatting of the page number or the page count, Ctrl+click or right-click the number, and then choose the format from the shortcut menu. For example, you can choose i, ii, iii or A, B, C instead of 1, 2, 3.

Controlling page breaks and section breaks

To make your documents appear the way you want, you'll need to make sure that your pages and sections break in the right places.

Using page breaks

You can tell Pages to insert a page break exactly where you want it by positioning the insertion point and choosing Insert ⇨ Page Break. A page break is normally invisible, but when you turn on the display of invisibles, it appears as a solid blue line with a page symbol at the right end (see figure 2.10).

2.10 The page-break symbol has the upper-right corner folded over.

You can delete a page break by clicking before it and then pressing Delete.

Genius

Instead of placing page breaks manually, you can use the Paragraph starts on a new page setting in the More pane of the Text Inspector to make Pages start a particular paragraph, or all paragraphs with a certain style, on a new page. See Chapter 3 for details.

Using section breaks

A *section* is simply a part of a document. A section can consist of a single page or multiple pages. Pages uses sections to control certain types of layouts, including headers and footers. As explained earlier in this chapter, you need to create different sections in a document if you want to have different headers and footers appear on the pages (other than having different headers and footers on the first page, or on left pages and right pages). Similarly, you need to use sections to create layouts that have different numbers of columns.

To create a new section, insert a section break by placing the insertion point and choosing Insert ⇨ Section Break. Like a page break, a section break is normally invisible, but you can show it by turning on the display of invisibles. Also like a page break, a section break appears as a solid blue line, but it has a section symbol at the right end — a sheet of paper with a filled-in column on the left (see figure 2.11).

2.11 A section break symbol looks similar to a page break symbol but it has a blue bar on the left and no corner is folded over.

You can delete a section break by clicking before it and then pressing Delete.

Making Your Documents Easier to Find with Spotlight

To make your documents easier to find with Spotlight, you can add metadata such as an author name, a title, keywords, and comments to them. To do so, click the Inspector button in the toolbar, click the Document Inspector button, and then work in the Info pane. Use the Range pop-up menu to choose between applying the metadata to the document as a whole (which is usually best) or just to the part of the document you've selected.

Adding Text to a Document

You can add text to a document by using any of the techniques discussed in Chapter 1 — for example, typing the text, pasting it in with or without formatting, or dictating it. But often you will also need to insert other types of text, such as hyperlinks or data fields from your Address Book. You may also need to format text quickly using keyboard shortcuts.

Genius

When you need to keep two words together so that Pages doesn't break a line between them, put a nonbreaking space between them by pressing Option+Space. This is good for proper names. When you display invisibles, a nonbreaking space appears as a dot with a caret over it.

Inserting hyperlinks

When you're distributing digital copies of a document rather than hard copies, it's often useful to include hyperlinks that the reader can click to open a Web site in her Web browser.

Here's how to insert a hyperlink in a Pages document:

1. **Choose Insert ⇨ Hyperlink, and then choose the type of hyperlink from the Hyperlink submenu: Webpage, Email Message, Bookmark, or Pages Document.** Pages inserts a standard hyperlink of the type you chose, opens the Inspector window if it's closed, and displays the Link Inspector with the Hyperlink pane displayed (see figure 2.12). For a hyperlink to a Pages document, Pages displays the Open dialog box so that you can pick the document.

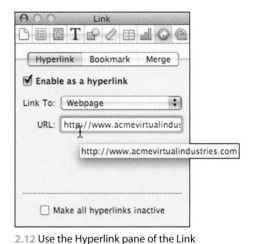

2.12 Use the Hyperlink pane of the Link Inspector to set up a hyperlink. To see the full URL, hold the mouse pointer over the URL box for a second.

2. **Type or paste the URL for the hyperlink in the URL box.**

3. **In your document, change the text displayed for the sample URL to the text you need.** For many URLs, you will want to display a description of the site rather than the URL itself.

4. **Close the Link Inspector.** Either click its red Close button or click the Inspector button on the Pages toolbar.

Replacing placeholder text

Most of Pages' templates include placeholder text, temporary text that's included to show you what the template looks like but which you're not meant to keep in the document. For example, many letter templates include placeholder text for the sender's address, the recipient's address, and the body of the letter, while page-layout templates such as those for newsletters include placeholder text for headings, the contents of text boxes, captions, and more.

You can replace most placeholder text by simply clicking to select the text, and then typing the text you want in its place. Where the placeholder text is in a text box, click the text box first, then click the placeholder and type the text. The text you type takes on the formatting applied to the placeholder text, so it has the right look for the template.

Inserting data from your Address Book or a Numbers document

If you need to send customized versions of the same document to different people, you can save time and effort by inserting data from your Address Book or from a Numbers document and performing a mail merge. You can merge into your document data from either a single group of your contacts (which is often useful, as you can create custom groups for different merges) or from your entire Address Book (which is occasionally useful — for example, if you're moving to a new address).

If the Pages template on which you based the document includes the merge fields you need, you're all set to merge in data. If not, you can add merge fields manually.

Whether you're going to merge data from your Address Book or from a Numbers document, start the Pages document as normal, and then type the standard text — the text that will be the same in each of the merged documents. Save and name the document as usual, and keep saving it as you make changes.

You can then add the merge fields as described next.

Adding Address Book fields to the document

Here's how to add Address Book fields to your document:

1. **Position the insertion point where you want to place the first field, choose Insert ⇨ Merge Field, and then pick the merge field from the submenu (see figure 2.13).**

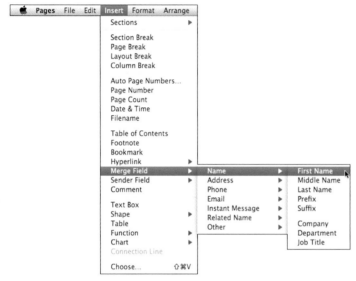

2.13 The Merge Field submenu contains further submenus of the fields from your Address Book.

2. **Pages opens the Link Inspector, displays the Merge pane, and inserts the field you chose (see figure 2.14).**

3. **Add other fields as necessary from the Insert ⇨ Merge Field submenu.**

2.14 You can insert Address Book fields quickly from the Merge pane of the Link Inspector.

Adding fields from a Numbers document

If your data source is in a Numbers document, follow these steps to choose the document and insert the fields:

1. **Click the Inspector button on the toolbar to open the Inspector window, and then click the Link Inspector button.** Click the Merge tab to display the Merge pane.

2. **Click the Choose button to display the Select a Mail Merge Source dialog box.**

3. **Select the Numbers Document option button.** Pages displays the Open dialog box.

4. **Select the Numbers document that contains the data, and then click Open.** The Select a Mail Merge Source dialog box shows the document you chose and the Table pop-up menu (see figure 2.15).

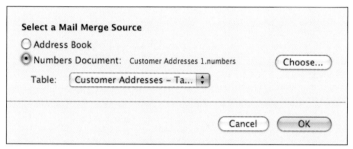

2.15 After choosing the Numbers document in the Select a Mail Merge Source dialog box, select the right table in the Table pop-up menu.

5. **In the Table pop-up menu, choose the table that contains the data.**

6. **Click OK.** Pages closes the Select a Mail Merge Source dialog box.

7. **Click the Add Field (+) button in the lower-left corner of the Merge pane in the Link Inspector and choose Add Merge Field from the pop-up menu.** Pages adds the First Name merge field to the list in the Merge pane and inserts it in the document.

8. **Click the arrow buttons on the First Name row in the Target Name column and choose the field you want from the pop-up menu (see figure 2.16).** (If you want to keep the First Name field, just leave it.) Pages changes the new First Name field in the list and in the document to the field you just chose.

9. **Repeat Steps 7 and 8 to add each of the other fields you need.**

Merging the data into the document

When you're ready to create the document, follow these steps:

1. **Choose Edit ⇨ Mail Merge.** Pages displays the Select a Mail Merge Source dialog box shown in figure 2.17.

2.16 Choose the field with which you want to replace the First Name field. Pages automatically adds it.

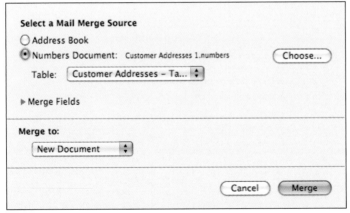

2.17 Choose the Address Book group or Numbers document table that contains the data to merge with your document.

2. **Make sure you've selected the right data source:**

● **Address Book.** With the Address Book option button selected, choose the group in the Address Book Group pop-up menu. Choose All if you want to include everyone in your Address Book.

● **Numbers Document.** With the Numbers Document option button selected, open the Table pop-up menu and choose the table that contains the data. If you need to use a different Numbers document, click the Choose button, and then select the document.

3. **Click the Merge Fields disclosure triangle to display the list of merge fields the document is using.** Figure 2.18 shows an example.

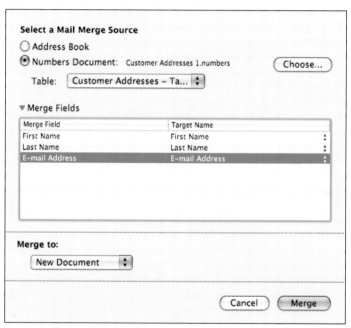

Select a Mail Merge Source

○ Address Book
● Numbers Document: Customer Addresses 1.numbers   (Choose...)

Table: [Customer Addresses – Ta... ◆]

▼ Merge Fields

Merge Field	Target Name
First Name	First Name
Last Name	Last Name
E-mail Address	E-mail Address

Merge to:
[New Document ◆]

(Cancel) (Merge)

2.18 To avoid surprises, check that your merge fields are correctly mapped to the data source before you perform the merge.

Note

If the message "Targets are missing, please select new targets" appears in red to the right of Merge Fields in the Select a Mail Merge Source dialog box, it means that one of the fields you've entered in the document isn't contained in the data source. This usually happens when you switch from one data source to another — for example, from your Address Book to a Numbers document. Expand the Merge Fields section of the dialog box and choose a different target for each field listed in red.

4. **In the Merge to pop-up menu, choose New Document if you want to create a new document.** This is usually the best choice, as it lets you check the data for omissions and errors. The alternative is to choose Send to Printer, which sends the merged data directly to the printer.

5. **If you're merging Address Book data, deselect the Substitute closest match check box if you don't want Pages to substitute data for any fields that are blank.** For example, if the work address is blank, Pages will substitute a home address (if there is one). The substitutions can be helpful, but you'd be wise to check the effects carefully with your data source.

6. **Click the Merge button.** Pages closes the dialog box and either creates a new document, which contains as many pages as are needed for a full copy of the document for each recipient, or sends the merged data to the printer.

If you merged the data to a document, check the document for errors, and then save it.

Genius

You can also merge data by clicking and dragging a card, multiple cards, or a group of addresses from Address Book to Pages. This is handy for merging a single address or for merging several addresses from different groups when you do not want to create a new group for the merge.

Changing your sender information for a document

If a document contains placeholders for the sender's name and address, Pages automatically fills them in using the data on the card named My Card in Address Book. If this is the data you want, you're set; if it is not, you can change it.

Note

You can go to your My Card card quickly in Address Book by choosing Card ⇨ Go to My Card. If that card is not the one you want to have as your My Card, choose the right card, and then choose Card ⇨ Make This My Card.

To change the card used for the sender data, open Address Book. Click the card you want to use and drag it onto the Address Book fields in the Pages document. Pages changes the fields to show the data from the new card.

Formatting text with keyboard shortcuts

Besides the common keyboard shortcuts discussed in Chapter 1 (such as ⌘+B for applying and removing boldface), Pages also provides keyboard shortcuts for formatting text with styles and adjusting list indentation. Table 2.1 has the details.

Table 2.1 Keyboard Shortcuts for Formatting Text in Pages

Keyboard Shortcut	Effect
⌘+Shift+T	Show or hide the Styles drawer
⌘+Option+C	Copy the paragraph style of selected text
⌘+Option+Shift+C	Copy the character style of selected text
⌘+Option+V	Paste the copied character or paragraph style
⌘+]	Increase list indentation
⌘+[	Decrease list indentation

You can also apply styles by pressing the F1 through F8 keys once you've assigned them to the styles you want to use. See Chapter 3 for details.

Genius

Adding New Template Pages to a Document

When you're working with a page-layout document, you can use the Pages pop-up menu on the toolbar to add new pages from the template to the document. The selection of pages depends on the document's template.

Genius

If you create a custom section in a document that you want to be able to reuse in other documents based on this template, you can add the section to the Pages pop-up menu. Click in the section, and then choose Format ➪ Advanced ➪ Capture Pages to display the Create Pages from the current section dialog box. Type a name for the page or pages, choose the page or pages in the Include pop-up menu, and then click OK.

Working with Microsoft Word Documents

The 800-pound gorilla of word processing is Microsoft Word, which runs on both Windows and the Mac (and, if you put in some effort, on Linux). As a smaller primate in the word-processing jungle, and one that runs only on the Mac, Pages needs to be able to import Word documents and export its own documents in the Word format.

Note

Pages can open documents in the Word 2007 (Windows) and Word 2008 (Mac) documents formats, which use the .docx file extension, as well as documents in the formats used by Word 2003 (Windows) and Word 2004 (Mac) and earlier versions, which use the .doc file extension.

To work with a Word document in Pages, simply open it as you would any other document: Press ⌘+O or choose File ➪ Open to display the Open dialog box, select the document, and click the Open button. From the Finder, you can Ctrl+click or right-click the document and choose Open With ➪ Pages from the shortcut menu.

Pages tries to convert all the document's contents to equivalents in Pages' own formats, which takes a few seconds (depending on how large and complex the document is).

Genius

Just to be clear — you can't open a Word document in Pages, edit it there, and then save your changes to the original Word document, as you might do using OpenOffice .org or another word processor. Instead, you have to save the document as a Pages document; if you want a Word version of that document, you can export a version of the document to Word.

Overall, Pages does a solid job of importing Word documents, importing all of the following items successfully:

- **Text paragraphs.** Pages can import pretty much any text paragraph that Microsoft Word can throw at it, including styles, tables, and lists.
- **Footnotes and endnotes.** Pages imports these correctly. You may run into problems with longer documents that contain many notes and ones where you've switched footnotes to endnotes, endnotes to footnotes, or each kind of note to the other.

- **Bookmarks, hyperlinks, and cross references.** Pages imports these correctly.

- **Tracked changes.** Pages imports the details of deleted text and inserted text, including the time and the user who made the changes. Because Pages itself marks only deleted text and inserted text, Pages marks any moved text in the Word document as a deletion in its old position (before you moved it) and an insertion in its new position. You'll find full details about using tracked changes in Chapter 5, including exchanging tracked changes with Microsoft Word.

- **Inline objects and floating objects.** Pages imports text boxes, graphics, charts, and other objects very impressively. For example, Pages can handle text boxes that are linked together in Word (so that text flows from one text box, when it's full, to the next).

But here are some of the things that don't come across so well:

- **Complex tables.** Pages smoothly handles straightforward tables — ones with a regular structure and no nested cells — but nested tables create problems. Table formulas can also be a problem: Pages successfully imports some table formulas, such as the SUM(ABOVE) and SUM(LEFT) formulas widely used in Word tables, but others produce an import warning saying "An unsupported field wasn't imported." You'll need to comb through imported tables and make sure everything's where it should be and that all the formulas are working.

- **Charts.** Pages imports Word charts with their underlying data, which is pretty impressive. However, you don't get any links to the data source (for example, a spreadsheet in Excel). It's a good idea to examine each chart closely to make sure that Pages has translated everything successfully.

- **Equations.** Pages simply removes Word's equations, which it calls "native equations."

These limitations may sound pretty negative — but if you're working with regular Word documents, such as business letters or reports, you'll probably find Pages' importing capabilities good enough.

Usually, the best way to proceed is to open the Word document in Pages and see what you get.

If Pages finds any problems in the Word document that it needs to deal with, it displays a dialog box telling you that some warnings occurred and inviting you to review them.

Usually, it's best to review the warnings right away, so that you know if you need to close the document, open it in Word and fix it, and then import it into Pages again. Click the Review button, and Pages displays the Document Warnings dialog box (see figure 2.19).

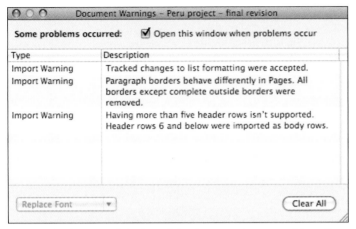

2.19 Use the Document Warnings dialog box to review any problems Pages encounters when importing a Word document.

If you prefer not to review the warnings now (for example, because you're facing a looming deadline), click Don't Review in the first dialog box. You can then pop up the list of warnings at any time by choosing View ⇨ Show Document Warnings from the menu bar.

Note

When you need to share a document you've created in Pages with someone who uses Word, save a copy as a Word document as discussed in Chapter 5.

How Can I Get the Most Out of Styles and Formatting?

Select the styles to import from:

Abseiling News – January

Header & Footer

HEADING 1

Heading 2

Heading 3

Month

TITLE

TITLE 2

(Select All)   (Deselect All)

☐ Replace duplicates   (Cancel)   (OK)

With fonts, colors, alignment, line spacing, indentation, background colors, and more, Pages gives you precise control of how the text in your document looks. With all these options, it's all too easy to spend ages perfecting your documents' formatting if you approach the process the wrong way. The key to formatting text efficiently is to use styles to format as much of your document as possible. You can then add special effects by applying direct formatting such as bold, italics, or colors where needed. A further benefit of using styles is that you can use outline view to develop the structure of your document quickly.

Formatting Text Quickly with Styles

A *style* is simply a collection of formatting settings that you can apply instantly to a paragraph or to one or more characters.

Pages provides four different kinds of styles:

- **Paragraph style.** A complete set of formatting for a paragraph. The paragraph style includes font formatting, alignment, indentation, and line spacing settings, and even a setting that controls which style Pages gives the next paragraph.

Genius

Each paragraph always has a paragraph style. In a new blank document, the first paragraph is in the Body style, so it appears as regular text. Rather than remove a paragraph style from a paragraph, you apply another paragraph style instead. By contrast, text doesn't usually have a character style, list style, or table of contents style unless it needs one or more of them, and you can remove these styles after applying them.

- **Character style.** A set of formatting for characters and words. For example, an Emphasis style may apply boldface to the selected text. You apply the character style "on top" of the paragraph style, adding extra character formatting to the formatting supplied by the paragraph style.

- **List style.** A set of formatting for creating numbered and bulleted lists. The list style adds the list formatting to the paragraph style rather than replacing the paragraph style.

- **Table of Contents style.** A set of formatting for table of contents entries — for example, so that the page number appears aligned right of the paragraph with tab leader dots leading up to it.

Applying styles

You can apply styles either from the Styles drawer or from the Paragraph Style button and Character Style button on the Format bar.

Opening and closing the Styles drawer

When you're applying styles to several paragraphs, the Styles drawer (see figure 3.1) is usually the easiest tool to use. You can open and close the Styles drawer any of these ways:

- Click the View button on the Toolbar and choose Show Styles Drawer or Hide Styles Drawer.

- Click the Styles Drawer button on the Format bar to open or close the Styles drawer.

- Choose View ⇨ Show Styles Drawer or View ⇨ Hide Styles Drawer from the menu bar.
- Press ⌘+Shift+T to open or close the Styles drawer.

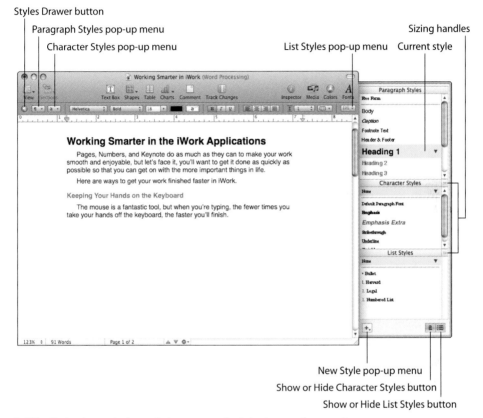

Styles Drawer button
Paragraph Styles pop-up menu
Character Styles pop-up menu
Sizing handles
List Styles pop-up menu   Current style

New Style pop-up menu
Show or Hide Character Styles button
Show or Hide List Styles button

3.1 The Styles drawer is the easiest way to apply styles to your documents.

The blue highlight in each of the panes shows which style is currently applied. For example, in the figure, Heading 1 is applied to the selected paragraph, with no character style and no list style applied on top of it.

You can resize the Character Styles area and the List Styles area of the Styles drawer by clicking and dragging the sizing handle at the right end of the bar. Or, if you don't need to use these styles, you can hide them by clicking the Show or Hide Character Styles button or the Show or Hide List Styles button.

The New Style pop-up menu in the lower-left corner of the Styles drawer lets you create new styles, as discussed later in this chapter.

Note

The Table of Contents styles appear in the Styles drawer only when you've selected a table of contents in the document.

Applying styles from the Styles drawer

Applying a style from the Styles drawer takes only a moment:

- To apply a paragraph style or list style to a paragraph, click in the paragraph, and then click the style name.

- To apply a paragraph style or list style to several paragraphs at once, select the paragraphs, and then click the style name.

- To apply a character style, select the text, and then click the style name.

Applying styles from the Format bar

When you've got the Format bar displayed, you can quickly apply a paragraph style from the Paragraph Style pop-up menu or a character style from the Character Style pop-up menu.

Applying styles using function keys

When you're typing, the quickest way to apply styles is via the function keys.

First, you need to set up the function keys you want to use. Follow these steps:

1. **Open the Styles drawer.** For example, press ⌘+Shift+T.

2. **Ctrl+click or right-click the style you want to assign a function key.**

3. **Highlight Hot Key on the shortcut menu, and then click the hot key you want.** You can use the function keys F1 through F8.

Once you've done this, simply press the function key to apply the style.

Genius

You may need to change your Mac's function key settings in order to apply styles using the function keys. Choose Apple menu ⇨ System Preferences, and then click Keyboard & Mouse to open the Keyboard & Mouse pane. Click the Keyboard tab, select the Use all F1, F2, etc. keys as standard function keys check box, and then quit System Preferences. You'll then need to press Fn with the function keys in order to use them for their special functions, such as volume control and brightness control.

Applying styles using copy and paste

When you've formatted text just the way you want it, you can copy the formatting to text elsewhere in the document, or to another document. Select the formatted text, open the Format menu, and choose Copy Paragraph Style or Copy Character Style, as appropriate. Then select the text you want to format (switch documents first if necessary), open the Format menu, and choose Paste Paragraph Style or Paste Character Style.

Genius

If you use the Copy Paragraph Style and Paste Paragraph Style commands often, customize the toolbar by adding the Copy Style button and Paste Style button to it. See Chapter 1 for instructions on customizing the toolbar.

Replacing one style with another style

You can quickly replace all instances of one style with another style. When you want to affect only the body text of the document, without affecting the text in text boxes and shapes, replace a style like this:

1. **Open the Styles drawer if it's closed.**

2. **Highlight the style you want to replace.**

3. **Click the arrow at the right of the style's name, and then choose Select All Uses of *Style* (where *Style* is the name of the style — for example, Select All Uses of Heading 1).** Pages selects all the body text that has the style applied.

4. **Click the style you want to apply instead of the current style.**

To replace all the instances of the style anywhere in the document, including those in text boxes and shapes, use Find and Replace instead. Follow these steps:

1. **Choose Edit ⇨ Find ⇨ Find to open the Find & Replace dialog box.**

2. **Click the Advanced tab to display the Advanced pane.**

3. **Delete any text from the Find box.**

4. **In the upper Style pop-up menu, choose the style you want to replace (see figure 3.2).**

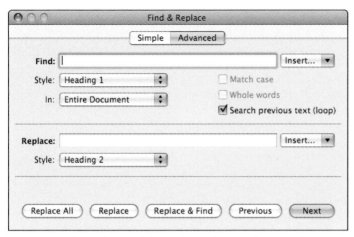

3.2 Use the Advanced pane of the Find & Replace window to replace one style with another in every part of the document.

5. **In the In pop-up menu, choose Entire Document.** (If you want to replace only in the body text, choose Main Text Body instead.)

6. **Delete any text from the Replace box.**

7. **In the lower Style pop-up menu, choose the replacement style.**

8. **Use the Replace buttons to replace the style as needed.** For example, click Replace & Find to replace the first instance and find the next, or click Replace All to replace all the instances of the style at once.

Applying basic formatting with paragraph styles

Paragraph styles provide the basic formatting for your documents, so start your formatting by applying the right paragraph style to each paragraph. For example, when writing a report, you may apply a Title style to the title, a Heading style to each heading, and a Quote style to each block quote.

To apply a paragraph style to a paragraph, simply click in the paragraph, and then choose the style from the Styles drawer or the Paragraph Style pop-up menu on the Format bar.

If there isn't a suitable style, either create one (as described later in this chapter) or apply the style that's closest in formatting and then tweak it by applying direct formatting (also described later in this chapter).

Genius

Each paragraph style includes a setting that controls the style of the next paragraph created when you press Return. For example, a Heading style is usually followed by a Body paragraph, while a Body paragraph is normally followed by another Body paragraph. You can override this setting by pressing Return before the end of the paragraph rather than right at the end.

To change the paragraph style, simply replace the current style by applying the style you want. You don't need to remove the existing paragraph style from the paragraph — in fact, you can't remove it.

Adding emphasis with character styles

Once you've applied the main formatting with paragraph styles, you can apply character styles to individual characters, words, or phrases to make them stand out from the rest of the text.

To apply a character style to a word, click in the word, and then choose the style from the Styles drawer or the Character Style pop-up menu on the Format bar.

To apply a character style to anything other than a word, select the text, and then apply the style.

Genius

Instead of using character styles, you can use direct formatting. For example, instead of applying the Emphasis character style that many Pages templates include, you can get a similar effect by applying boldface. But by using the style rather than direct formatting, you can keep your formatting consistent. And if you update the style, Pages automatically applies the changes for you throughout the document.

To remove a character style, click in the word or select the text that bears the style, and then click None at the top of the Character Styles pane in the Styles drawer or open the Character Styles pop-up menu and choose None.

Creating lists with list styles

To create a numbered or bulleted list in a document, you apply a list style to the paragraph or paragraphs. The list style goes on top of the paragraph style, adding the list formatting to the paragraph style rather than replacing it.

To apply a list style to a paragraph, click in the paragraph, and then click the style in the List Styles pane in the Styles drawer. Alternatively, open the List Styles pop-up menu on the Format bar and click the list style.

Genius If you're familiar with Microsoft Word, you'll find Pages' list styles substantially different from Word's. In Word, a list style is a paragraph style, so when you apply a list style, it replaces the paragraph style that was previously applied. In Pages, by contrast, the list style adds formatting to the paragraph style rather than replaces it.

To apply a list style to multiple paragraphs, select the paragraphs, and then click the style in the List Styles pane or the List Styles pop-up menu.

To remove a list style, click in the paragraph or select the paragraphs that use the style, and then click None at the top of the List Styles pane or the List Styles pop-up menu.

Importing styles from another Pages document

If you have the styles you want, but they're in another Pages document, you can quickly import them. Follow these steps:

1. **Open the document into which you want to import the styles.**

2. **Choose Format ⇨ Import Styles to display the Open dialog box.**

3. **Select the document that contains the styles, and then click Open to display the Select the styles to import from dialog box (see figure 3.3).**

4. **Select the styles you want to import:**

   ● Click Select All to select all the styles, or click Deselect All to deselect the ones you've selected.

   ● Click a style, and then Shift+click another style to select all the styles in between.

   ● ⌘+click a style to add it to your selection.

5. **Choose how to handle styles you import that have the same names as styles in the open document:**

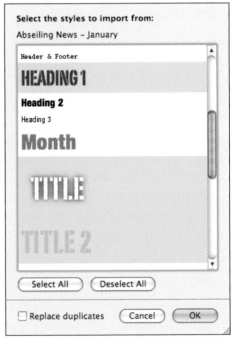

3.3 The Select the styles to import from dialog box lets you quickly pull in styles from another Pages document.

- Select the Replace duplicates check box if you want to simply overwrite the existing styles with the ones you import. Text using the existing styles takes on the formatting of the imported styles. Do this when you want to switch the document to the new styles.

- Deselect the Replace duplicates check box if you want to keep your existing styles. Pages then adds a number to the names of the imported styles to distinguish them; for example, renaming the Caption style to Caption 2. You can then apply the styles manually as needed. Do this when you want to make the styles available in the document without changing its current formatting.

6. **Click OK to import the styles.** You can then use the styles as usual from the Format bar and the Styles drawer.

Importing styles from a Microsoft Word document

If you or your colleagues use Microsoft Word, you'll probably find Pages' capability to import the styles from Word documents a great time-saver.

Here's how this works: You simply open the Word document in Pages, and you get all the document content — text and so on — plus each style used in the document.

Genius

Pages doesn't import Word styles that aren't used in the document, which saves you from having a huge list of styles you probably didn't need (for example, the dozens of list styles and table styles most Word documents have). So if you need to have a certain style available in the Pages document, make sure it's used in the Word document before you import the document. Alternatively, you can paste in the style from another Word document later.

Just open the document as usual from Pages. Choose File ⇨ Open to display the Open dialog box, select the Word document, and then click Open. If you're working from the Finder, Ctrl+click or right-click the document, and then choose Open With ⇨ Pages from the shortcut menu.

When Pages opens the document, it displays the Word styles. You can then apply them from the Styles drawer or the Paragraph Styles pop-up menu and Character Styles pop-up menu as usual. You can also assign hot keys for styles, as discussed earlier in this chapter.

Adding a Few Styles from a Word Document

Sometimes you may need to add just a few styles to your Pages document from another Word document. If you have Word on your Mac, you don't need to open the document in Pages and import all the styles used in the document. Instead, follow these steps:

1. **Open the document in Word.**

2. **Copy one or more paragraphs that have the style or styles applied.**

3. **Switch to Pages, and then paste in what you copied.** Pages picks up the styles from the text.

You can now delete the pasted text if you don't need it. Pages retains the styles, and you can apply them from the Styles drawer or from the Format bar.

Creating Custom Styles

When the styles in Pages' templates don't meet your needs, and you can't bring in the style you want from another document, you can create custom styles of your own.

Creating your own styles in Pages is very easy. You simply set up a paragraph or some text the way you want it to appear and then create a style from it.

The fastest and easiest way to create a new style is by changing one of the existing styles to make it look the way you want it. So start by typing a paragraph of text — or clicking in an existing paragraph — and applying the style that's nearest to what you need.

Genius

You can also create a new paragraph style from scratch by applying the Free Form style to the paragraph before you start customizing the style. But usually you can save time and effort by starting from one of the existing styles.

The following sections explain the main types of changes you can make to the style.

Changing the font formatting

You can change the font formatting of the text either by using the controls on the Format bar or by clicking the Fonts button on the Toolbar and using the Font panel. Which works better depends on the way you prefer to work, but generally, the Format bar controls are better for quick changes to the most widely used types of formatting, while the Font panel lets you make more extensive changes and gives you access to the more specialized types of formatting.

Genius

You can open the Font panel by pressing ⌘+T or choosing Format ⇨ Fonts ⇨ Show Fonts.

You can make these main changes in the Fonts panel (see figure 3.4):

- **Font.** Choose a font collection in the Collections box (or simply choose All Fonts), and then choose the font family in the Family box. In the Typeface box, choose the typeface — for example, Regular, Italic, Bold, or Bold Italic. Then choose the Size in the Size box, or drag the Size slider beside the Size list.

- **Underline and Strikethrough.** Use the Text Underline pop-up menu and the Text Strikethrough pop-up menu if you need to apply these effects.

- **Text Color and Document Color.** Click the appropriate button to open the Colors panel, pick the color, and then click the Close button (the red button) to return to the Fonts panel.

- **Text Shadow.** Click the Text Shadow button to toggle text shadow on. You can then use the Shadow Opacity slider, Shadow Blur slider, Shadow Offset slider, and Shadow Angle knob to adjust the shadow.

- **Font Panel Actions.** This pop-up menu gives you access to options for customizing the Font panel and for choosing other font options. For example, you can choose Show Preview to display a font preview at the top of the Font panel, or click Typography to open the Typography window, which lets you change items such as ligatures, cursive connections, and text spacing.

If you're creating a character style, you're now ready to create it, as described later in this chapter. If you're creating a paragraph style, list style, or table of contents (TOC) style, you'll usually need to make further formatting changes, as discussed next.

Text Underline pop-up menu
Text Strikethrough pop-up menu
Document Color button   Shadow Opacity slider   Shadow Offset slider
Text Color button   Text Shadow button   Shadow Blur slider   Shadow Angle knob

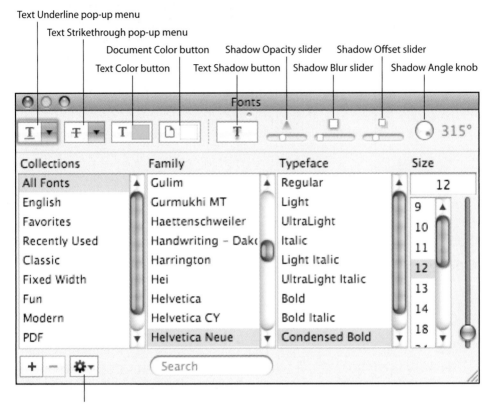

Font Panel Actions pop-up menu

3.4 The Fonts panel lets you change everything from the font, size, and color to text shadows.

Changing the paragraph formatting

You can change alignment and line spacing quickly by using the buttons on the Format bar, but to change other paragraph formatting, you need to use the Text Inspector. Click the Inspector button on the Toolbar to open the Inspector window, and then click the Text Inspector button.

Changing the text formatting

Start by clicking the Text tab to display the Text pane (if it's not already displayed). Figure 3.5 shows the Text pane of the Text Inspector. Here, you can change the following types of formatting:

- **Color.** Click the button to display the Colors window, then choose the color and close the window. If you've already set the color using the Fonts panel, you won't need to set it again.

- **Alignment.** Choose the horizontal alignment: Align Left, Center, Align Right, or Justify. For a table cell, you can choose the fifth alignment button to left-align text but right-align numbers. You can also choose the vertical alignment (Align Top, Align Middle, or Align Bottom).

● **Spacing.** Drag the Character slider, Line slider, Before Paragraph slider, After Paragraph slider, and Inset Margin slider to set the spacing you want. Alternatively, use the spin boxes. The Inset Margin measurement is the distance between the edge of an object and the beginning of the text.

Genius

To set the spacing between two paragraphs, Pages compares the After Paragraph spacing of the first paragraph with the Before Paragraph spacing of the second paragraph, and applies the larger of the two measurements. It doesn't add the two measurements together (unlike Microsoft Word and various other applications).

Justify button

Align Right button | Align Text Left and Numbers Right button

Center button | Align Top button

Align Left button | Align Middle button

Text Color button | Align Bottom button

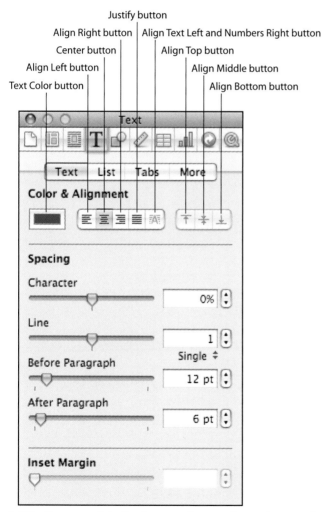

3.5 The Text pane of the Text Inspector lets you change the text color and alignment, spacing between paragraphs, and inset margin.

93

Changing the list formatting

If you're creating a list style, click the List tab in the Text Inspector to display the List pane (see figure 3.6). You can then set up the exact indents and bullets or numbering you need like this:

1. **Set the indent level.** You can either type a number between 1 and 9 (inclusive) in the Indent Level box, or click the Decrease Indent Level button or Increase Indent Level button to change the level in the box.

Increase Indent Level button

Decrease Indent Level button

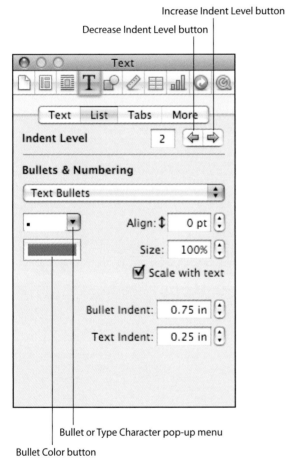

Bullet or Type Character pop-up menu

Bullet Color button

3.6 When you're creating a list style, set up indents and bullets or numbering in the List pane of the Text Inspector.

2. **Choose the bullet or numbering.** Open the Bullets & Numbering pop-up menu, and then choose the type of bullet or numbering:

- **No Bullets.** Choose this option to remove bullets from a list.

- **Text Bullets.** Choose this option for a variety of text-based bullets, from standard round ones to asterisks and crosses, to check marks and arrows.

- **Image Bullets.** Choose this option to pick from a selection of graphical bullets, including check boxes, sun symbols, and push-pins.

- **Custom Image.** Choose this option to pick an image file on your Mac to use as a bullet. Mac OS X automatically creates a miniature version of the image.

- **Numbers.** Choose this option to create a single-level numbered list.

- **Tiered Numbers.** Choose this option to create a multilevel numbered list.

3. **Choose the details and positioning of the bullet or numbering.** The Bullets & Numbering section of the List pane shows different controls depending on your choice in Step 2. For example:

- **Bulleted list.** Choose the bullet type or image. Use the Align box if you need to move the bullet up or down relative to the text. Use the Size box if you want to increase or decrease the bullet's size. Select the Scale with text check box if you want the bullet's size to change when you adjust the text's size. (This is usually a good idea for both text bullets and graphical bullets.) Use the Bullet Indent box to set the distance of the bullet from the left margin, and use the Text Indent box to set the distance of the text from the bullet.

- **Single-level numbered list.** Figure 3.7 shows the available controls. Open the Number Style pop-up menu and choose the number format you want — for example, 1. 2. 3. 4. or A) B) C) D). If you need to continue the list's numbering from the

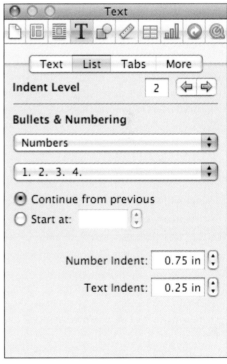

3.7 When creating a numbered list, choose whether to continue the numbering from the previous list or start it afresh.

previous list, select the Continue from previous option button; otherwise, select the Start at option button and type the starting number in the box. Use the Number Indent box to set the distance of the number from the left margin, and use the Text Indent box to set the distance of the text from the number.

- **Multilevel numbered list.** Follow the instructions in the previous paragraph to choose the numbering for the first level of the list. Now click the Increase Indent Level button to move to the next level, and then choose the numbering and number positioning for this level. Repeat the process for each level of the list.

Changing the tab formatting

If you use tabs to lay out complex text, you'll often need to adjust the document's tabs to suit the layout. Pages comes set with left-aligned tabs at half-inch intervals, but you may need to use different spacing, different tab alignments (Pages also offers centered tabs, right-aligned tabs, and tabs aligned on the decimal point in numbers), or add tab leader characters (such as dots) for a table of contents.

To set tab formatting in a style, use the Tabs pane in the Text Inspector. Click the Tabs tab in the Text Inspector to display the pane (see figure 3.8).

First, set the indents for the paragraph in the First Line box, Left box, and Right box in the Paragraph Indents area. The First Line indent is independent of the Left indent, so you can produce a hanging indent (or "outdent," as some people call it) by setting a lower value in the First Line box than in the Left box.

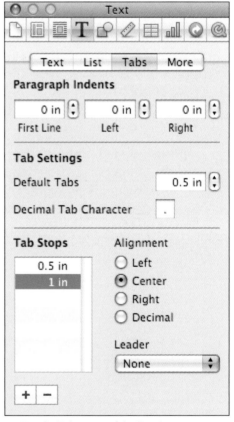

3.8 Use the Tabs pane of the Text Inspector to equip the style with tabs that have the alignments and positions you need.

Now move on to the Tab Settings area, and set the standard interval between tabs in the Default Tabs box. You can also change the character used for aligning the text on decimal tabs by typing in another character in the Decimal Tab Character box, but the period is normally the best choice.

Now use the controls in the Tab Stops area to set the tabs you need:

1. **Click the New Tab (+) button to add a new tab to the Tab Stops box.** Pages uses the spacing you set in the Default Tabs box.

2. **To change the tab position, double-click the new tab in the Tab Stops box, and then type the position you want.** For example, type **1.5 in** to set a tab at 1.5 inches on the ruler.

3. **Choose the alignment for the tab.** Select the Left option button, Center option button, Right option button, or Decimal option button in the Alignment area.

4. **If the tab needs a tab leader, open the Leader pop-up menu and choose dashes, dots, or an underline, as needed.**

To delete a tab, click it in the Tab Stops box, and then click the Delete Tab button (the – button in the lower-left corner of the Tabs pane).

Genius Tabs are great for laying out short pieces of columnar text, but if you have multiple lines for each entry, keeping text aligned with tabs can become a nightmare. In this case, using a table will usually save you plenty of time and effort. If you need columns of flowing text, use columns instead.

Changing borders, fills, pagination, and further options

Now click the More tab to display the More pane (see figure 3.9). Here you can set the following options for the paragraph:

- **Border & Rules.** Choose the line style in the Line Style pop-up menu, a color from the color picker, and then a weight in points. Specify where to place the line by clicking the Line Above button, Line Below button, Line Above and Below button, or Border button. Set the distance in points from the text to the border in the Offset box.

Line Style pop-up menu

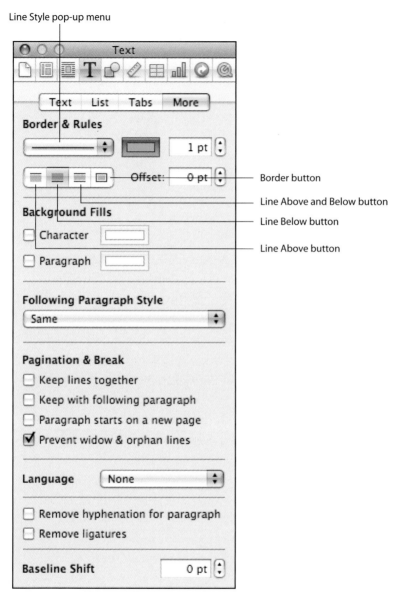

Border button

Line Above and Below button

Line Below button

Line Above button

3.9 The More pane in the Text Inspector lets you add borders, rules, and background fills; control the following paragraph's style, pagination, and breaks; and set the language for the style.

- **Background Fills.** To apply a background fill, select the Character check box or the Paragraph check box, and choose the color in the Color window that Pages displays.

- **Following Paragraph Style.** In this pop-up menu, choose the name of the style you want to use for the next paragraph. For example, after a Subheading style, you might choose a style such as Body Text or Body Text First (a style for the first body paragraph, which might then have Body Text as its following-paragraph style). Choose Same if you want to use the same style, as you normally would for body text styles.

Genius

The Following Paragraph Style setting in the More pane of the Text Inspector can save you a huge amount of time. When you choose a suitable following-paragraph style for each style, Pages automatically applies many of the styles you need, and you do not need to apply them yourself.

- **Pagination & Break.** These settings enable you to keep all the lines of a paragraph together, keep a paragraph with the paragraph that follows it, start a paragraph on the next page, and prevent typeset *widow lines* and *orphan lines*. See the in-depth discussion toward the end of the chapter for more on these features.

- **Language.** If the style always uses the same language, choose it in the Language pop-up menu to make sure the spelling checker uses the right dictionary. If the style uses two or more languages, choose None.

- **Hyphenation and ligatures.** Select the Remove hyphenation from paragraph check box if you want to suppress hyphenation in the style. Select the Remove ligatures check box if you want to remove typeset ligatures (such as œ) from the style.

- **Baseline Shift.** If you want to move the baseline of the text up or down, set the number of points in this box. The *baseline* is the imaginary line on which the bottoms of letters sit; the descenders (down strokes) on letters such as *g* and *y* go below the baseline.

Creating and naming your new style

Once you've set up your sample paragraph or text with all the formatting needed for your new style, you are ready to create the style:

1. **Click in your sample paragraph or select your sample text.**

2. **Click and hold down the New button at the bottom of the Styles drawer to produce a pop-up menu showing the different style types.**

3. **Choose Create New Paragraph Style from Selection, Create New Character Style from Selection, or Create New List Style from Selection, as appropriate.** Pages displays the New Paragraph Style, New Character, or New List Style dialog box. Figure 3.10 shows the New Character Style dialog box with a name already entered. The New Paragraph Style and the New List Style dialog boxes are the same, except that they don't have the Include all character attributes disclosure triangle.

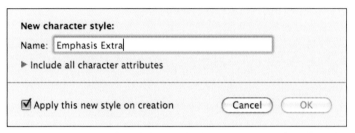

3.10 Name your new style and decide whether to apply it to the selected text.

4. **Type the name for the style in the Name box.**

Genius

When you create more than a few styles, use a naming convention so that related styles appear next to each other in the Styles drawer. For example, using the names "Body Subtle" and "Body Dramatic" keeps the names together, whereas "Subtle Body" and "Dynamic Body" separates them by a wide span.

5. **When creating a character style, choose whether to include all the character attributes of the selection or just some of them.** Follow these steps:

   1. Click the Include these character attributes disclosure triangle to open the hidden section of the dialog box (see figure 3.11). You can now see all the details of the character formatting: the font, size, character spacing, bold italic, color, and much more.

   2. Deselect the check boxes for the attributes you don't want to include. Normally, the easiest thing is to click the Select Overrides button. This makes Pages select the check boxes only for the character attributes that the formatting overrides in the style. Usually, this is the extra formatting that you want the style to apply. You can also deselect and select check boxes manually or use the Select All and Deselect All buttons.

6. **Deselect the Apply this new style on creation check box if you don't want to apply the style to the selected text.** (Pages selects this check box automatically.) When you deselect this check box, the text retains its current styles, even though its formatting is that of the new style you're creating.

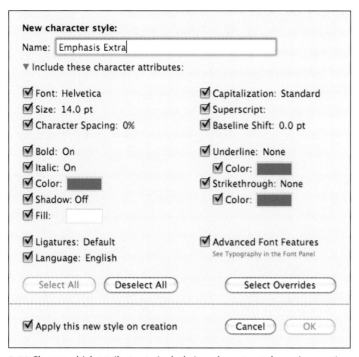

New character style:

Name: Emphasis Extra|

▼ Include these character attributes:

☑ Font: Helvetica
☑ Size: 14.0 pt
☑ Character Spacing: 0%

☑ Capitalization: Standard
☑ Superscript:
☑ Baseline Shift: 0.0 pt

☑ Bold: On
☑ Italic: On
☑ Color: ▩
☑ Shadow: Off
☑ Fill: ☐

☑ Underline: None
☑ Color: ▩
☑ Strikethrough: None
☑ Color: ▩

☑ Ligatures: Default
☑ Language: English

☑ Advanced Font Features
See Typography in the Font Panel

(Select All) (Deselect All)   (Select Overrides)

☑ Apply this new style on creation   (Cancel) (OK)

3.11 Choose which attributes to include in a character style you're creating.

7. **Click OK to close the dialog box.** Pages creates the style and adds it to the Styles drawer and the pop-up menus.

Genius When you're creating a paragraph style, you can also use the Create New Paragraph Style from Selection command on the Format menu or on the shortcut menu for text (which you can display by Ctrl+clicking or right-clicking in a paragraph).

Delete a style from a document

When you no longer need a style in a document, you can delete it. This works both for built-in styles and custom styles.

Genius If there's even a remote possibility that you may need the style again, just leave it in the document rather than deleting it. A style is just a small collection of formatting information, so it takes up only a tiny amount of space.

Here's how to delete a style:

1. **Open the Styles drawer.** For example, press ⌘+Shift+T or click the Styles Drawer button on the Format bar.

2. **Highlight the style in the Styles drawer, click the disclosure triangle to display the menu, and then click the Delete Style button.** If the style is used in the document, Pages warns you (see figure 3.12) and prompts you to choose the style with which to replace it.

The style you want to delete is used in the document. Choose a style to replace it:

Body

Cancel   Replace

3.12 Before deleting a style that's used in the document, you must choose the style with which to replace it.

3. **Choose the style in the pop-up menu, and then click the Replace button.** Pages replaces all instances of the style, and then deletes the style from the document.

Applying direct formatting to text, paragraphs, and elements

Besides applying styles as discussed earlier in this chapter, you can apply direct formatting to text, paragraphs, or other elements. For example, you may change the alignment or margins of a paragraph, the font size or weight, or the tabs positions. Pages calls this *overriding* a style or *applying a style override*.

Applying a style override

To apply a style override, use the techniques discussed earlier in the chapter. Use the controls on the Format bar to make quick changes, and open the Font panel or the Text Inspector when you need to make extensive changes or access less widely used options.

Genius

To keep your formatting fast and consistent, apply style overrides as seldom as possible. If you need to apply the same direct formatting to several items, create a style for it. You can then update the style, and Pages automatically applies the update to all the text that has the style. You can also copy the style to another document if you need it there.

Removing a style override

To remove a style override, follow these steps:

1. **Select the text or paragraph that contains the style override.**

Genius

If you have the Styles drawer open, you can instantly tell when there's a style override applied to the current style. When there is a style override, Pages displays the disclosure triangle for the style in red rather than the typical gray.

2. **Open the Styles drawer if it's closed. For example, click the Styles Drawer button on the Format bar.**

3. **Click the red disclosure triangle, and then choose Revert to Defined Style from the pop-up menu.**

Keeping Paragraphs and Lines Together

Normally, Pages flows your text automatically from the end of one page of a word-processing document to the beginning of the next page, so you don't need to organize the page breaks yourself. But sometimes you may need to make a particular part of a document start on a new page or force certain paragraphs or lines to stay together rather than breaking across pages.

Starting a paragraph after a page break

When you want a paragraph to start on a new page, you can simply insert a page break before it by positioning the insertion point and choosing Insert ➪ Page Break. If you've turned on the display of invisibles (for example, by choosing View ➪ Show Invisibles or pressing ⌘+Shift+I), the page break appears as a blue line with a page symbol at the right end (see figure 3.13).

Working·Smarter·in·the·iWork·Applications¶

Pages,·Numbers,·and·Keynote·do·as·much·as·they·can·to·make·your·work· smooth·and·enjoyable,·but·let's·face·it,·you'll·want·to·get·it·done·as·quickly·as· possible·so·that·you·can·get·on·with·the·more·important·things·in·life.¶

Here·are·ways·to·get·your·work·finished·faster·in·iWork.¶

3.13 A manual page break appears as a heavy blue line with a page symbol.

However, it's often handier to set Pages to start a new page automatically for the paragraph like this:

1. **Click anywhere in the paragraph.**

2. **Click the Inspector button on the toolbar to open the Inspector window.**

3. **Click the Text Inspector button, and then click the More tab to display the More pane (shown in figure 3.9, earlier in this chapter).**

4. **Select the Paragraph starts on a new page check box.** Pages moves the paragraph down so that it starts on a new page.

5. **Close the Inspector window unless you need to keep it open.** For example, you may need to adjust the settings for other paragraphs.

Keeping the lines of a paragraph together

When you have a very short paragraph — one that contains two or three lines — at the end of the page, it can run afoul of either or both of two typesetting rules:

- **Orphan lines.** An *orphan line* is a paragraph's first line that appears alone at the bottom of a page.

- **Widow lines.** A *widow line* is a paragraph's last line that appears alone at the top of a page.

Traditionally, typesetters and proofreaders have considered orphan lines and widow lines undesirable, because they can look like standalone one-line paragraphs and thus make the pages harder to read. So — like most other word-processing applications — Pages comes set to prevent orphan lines and widow lines by default. To prevent them, Pages simply bumps a two- or three-line paragraph from the end of one page to the top of the next page rather than breaking it.

If you don't mind widow and orphan lines, you can stop Pages from preventing them. Simply deselect the Prevent widow & orphan lines check box in the More pane of the Text Inspector.

Other times, you may need to keep longer paragraphs together so that they don't break from one page to the next. For example, in complex contracts that have several layers of numbering (1.a.i, 1.a.ii, and so on), it's often desirable to prevent paragraphs from breaking across pages.

To keep all the lines of a paragraph together, click in the paragraph, and then select the Keep lines together check box on the More tab of the Text Inspector pane.

Keeping two or more paragraphs together

When laying out a document, you'll often need to keep two paragraphs together rather than let-ting the first appear at the bottom of one page and the second at the top of the next page. For example, rather than strand a heading (a "paragraph" with a heading style applied to it) at the bot-tom of a page without a body text paragraph after it, it's best to move the heading to the next page so that it's with the text that follows it. Similarly, you may want to make sure that a caption appears with its figure or table rather than on the next page.

To keep a paragraph with the next, click in the paragraph, and then select the Keep with following paragraph check box on the More tab of the Text Inspector pane.

Caution

It's best to use the Keep with following paragraph setting only for a couple of para-graphs at a time. While you *can* set a whole sequence of paragraphs to stay together, be careful — you may end up with a short page because Pages has moved a large block of text to the next page. For this reason, it's usually best not to apply this set-ting to body text styles, because it's normal to have several body text paragraphs in sequence.

You can override the Keep with following paragraph setting by inserting a manual page break between paragraphs that have this setting applied. Just choose Insert ⇨ Page Break to insert the page break.

Creating Your Own Templates

Pages includes a great selection of word-processing and page-layout templates to give you a jump-start on creating many different kinds of documents, but you can save even more time by creating your own templates that contain exactly the styles and boilerplate text you need.

As discussed in Chapter 1, creating a template could hardly be easier: You simply create a docu-ment that contains the text, objects, styles, and layout you need; choose File ⇨ Save As Template; and then save the template in the My Templates folder. The template then appears along with Pages' built-in templates in the Template Chooser window, so you can quickly create a new docu-ment based on it.

As usual, the devil is in the details of what you put in the template. Bear these five suggestions in mind when planning your templates:

- **Jot down exactly what your template will do and who will use it.** This will help you to include everything the template needs and make it as useful as possible. At this planning stage, you may realize that you need to create two (or more) related templates rather than just one.

- **Base your templates on Pages' built-in templates wherever feasible.** There's no point in reinventing the wheel, and you can often save time by adapting an existing template rather than starting from scratch. For example, if you're creating a template for a business letter, see if you can use one of Pages' letter templates as the basis. Having the merge fields already in place will save your having to insert them again.

- **Make your template as close to the final document as possible.** This may seem to be stating the obvious, but you'd be amazed how many people go only partway with templates. For example, say you're creating templates to help reps respond quickly and accurately to customer-service queries. Having a separate template for each major topic helps the reps create the letters faster and more easily than one general-purpose template in which they need to enter more text for each letter. (You can also create Auto-Correction entries to help your colleagues enter essential information in documents quickly and accurately.)

Genius

Normally a template contains *less* material than the final document requires — but you can frequently save time by creating a template that contains *more* material than the final document. For example, you can create a template that contains three paragraphs of boilerplate text, and then delete those the document doesn't need. Deleting extra material is often faster and easier than inserting new material.

- **Create a sequence of related templates.** Say you produce several different newsletters for your company. You could create a single template and then change the masthead and taglines along with the content for each newsletter, but you could probably save time and effort by creating a separate template for each newsletter.

- **Exploit styles to the fullest.** As discussed earlier in this chapter, use the following paragraph style setting to help ensure that the person creating the document needs to apply styles manually as little as possible.

Developing a Document's Structure with Outline View

When you need to quickly create the structure of a word-processing document, use Pages' outline view. This view lets you quickly set down the headings and subheadings for the document and shuffle them into the right order.

Note

Outline view isn't available in page-layout documents because they don't have a suitable flow of text.

Open a word-processing document (or create and save a new document), and then switch to outline view by clicking the Outline button on the toolbar or choosing View⇨Show Document Outline. Pages displays the document as an outline of headings indented to different levels.

If the document already contains headings, Pages displays them all; each heading has a diamond to its left, and each body paragraph has a blue line, as you see in figure 3.14. A blue diamond indicates a heading with no collapsed headings or body paragraphs under it, while a white diamond indicates that there are hidden headings, body paragraphs, or both.

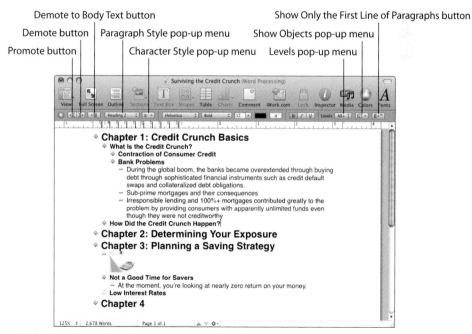

3.14 Outline view displays the document as collapsible headings.

If the document is new, Pages automatically applies the Heading 1 style to the first paragraph so that you're ready to start creating the outline.

You can now work on the outline like this:

- **Add a heading.** With the insertion point at the end of a paragraph, press Return. Pages creates the new paragraph at the same heading level as the previous paragraph, so you can just keep typing.

- **Add a subheading.** Press Return to create a new paragraph, and then press Tab or click the Demote button to move down to the next heading level. For example, pressing Tab from Heading 1 moves you down to Heading 2; pressing Tab again gives you Heading 3. Type the subheading. You can then press Return to create another subheading at the same level.

- **Demote an existing heading.** Click anywhere in the heading and press Tab or click the Demote button.

- **Promote an existing heading.** Click anywhere in the heading and press Shift+Tab or click the Promote button.

Genius

You can also promote or demote a heading by clicking and dragging its blue diamond to the left or right. Pages displays a guideline to show the level you've reached. Usually, though, it's easier to press Shift+Tab or Tab or to click the Promote or Demote button on the toolbar.

- **Expand or collapse a heading section.** Double-click a blue diamond to display the headings under it (if they're currently hidden) or to hide them (if they're currently displayed).

- **Select a heading section.** Click the blue diamond to select a heading and all the subheadings and text under it.

- **Move a heading or heading section.** Click in a heading, or select a section. Then click and drag the blue diamond up or down the outline. Pages displays a guideline to show you where the item you're dragging will land.

- **Change the heading levels displayed.** Open the Levels pop-up menu on the Format bar and choose how many heading levels to display. You can choose from 1 to 9 heading levels; choose All if you want to see the body text as well.

Note The Levels pop-up menu changes its readout to show how many levels are displayed. For example, 3+ means that levels 1, 2, and 3 are displayed, but you've expanded or collapsed some of them; All+ means that all levels are displayed, but you've collapsed some of them.

- **Choose whether to show inline pictures and movies actual size or as thumbnails.** Open the Show Objects pop-up menu and choose Actual or Thumbnail. Viewing the objects as thumbnails is usually helpful, as it lets you see more of your document at once.

- **Show only the first line of each paragraph.** Click the Show Only the First Line of Paragraphs button to switch between showing each paragraph in full and showing only its first line. This setting is especially helpful when you've displayed body paragraphs — by reducing them to their first lines, you can see more of the document as a whole.

Genius To make the most of outline view, don't take the heading levels too literally. Almost no final documents actually use nine levels of headings, because that complex a hierarchy confuses the reader. But you can benefit from using as many of the nine levels as you need when creating your documents. Simply change the lower levels to body text (or other elements) as appropriate when you've put everything in order.

You can switch back to normal view at any time by clicking the Outline button on the toolbar again or choosing View ⇨ Hide Document Outline.

Note When you save a document in outline view and then close it, the next time you open it, Pages automatically displays it in outline view. If you've been working in outline view but want the document to open normally, switch out of outline view before you save it and close it.

109

What Special Formatting Can I Use to Give My Documents Impact?

Creating content for your documents is half the battle; to get maximum impact you'll need to format your documents so they are as powerful and persuasive as possible. This chapter shows you how to use tabs, tables, and columns to lay out your documents; how to add backgrounds, shapes, charts, and images to make the documents look great; and how to position text precisely where you want it by flowing it through a series of linked text boxes.

Using Tabs

When you need to space out columns of text at precise intervals, you can use tabs. As in most word-processing applications, you set tab stops in the paragraph formatting to tell Pages where you want the tabs, and then press the Tab key to insert a tab that takes you to the next tab stop.

Whereas typewriters provide a single kind of tab, Pages provides four kinds:

- **Left.** This is the standard type of tab, just like the one on a typewriter. Text is aligned left at the tab.

- **Center.** Text is centered on the tab, so when you type text, it moves equal distances to the left and right of the tab. A center tab is useful for display text that's only part of a paragraph — for example, a centered part of a header or footer.

- **Right.** The opposite of a left tab. Text is aligned right at the tab and moves to the left as you type. Right tabs are great for positioning text so that it ends at the right margin, which is often useful in headers and footers.

- **Decimal.** Text is aligned at the decimal point. Decimal tabs are good for aligning columns of figures that have different numbers of decimal places.

Figure 4.1 shows an example of each kind of tab.

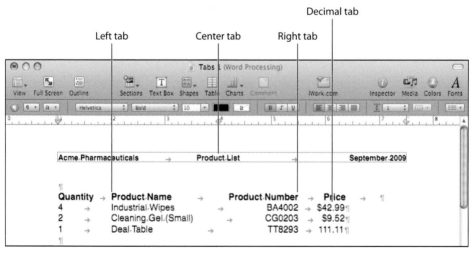

4.1 Pages gives you right tabs, center tabs, and decimal tabs as well as regular left tabs.

Setting a tab

The quick way to set tabs is to use the horizontal ruler like this:

1. **Display the horizontal ruler if it's hidden.** Press ⌘+R or click the View pop-up menu button on the toolbar and choose Show Rulers to display the ruler. You can also choose View ➪ Show Rulers from the menu bar.

2. **Click in the horizontal ruler where you want to place the tab.** Pages places a left tab there.

3. **If you want a different kind of tab, Ctrl+click or right-click the tab and choose the type you want from the shortcut menu (see figure 4.2).**

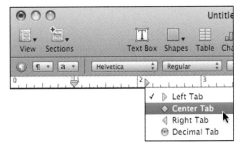

4.2 Click in the horizontal ruler to place a tab, then Ctrl+click or right-click it and change the type if necessary.

Moving and removing tabs

To move a tab to another tick mark, click and drag it along the horizontal ruler to where you want it. To move a tab freely to exactly where you want it, ⌘+click and drag it.

To remove a tab, click and drag it off the horizontal ruler. Drop the tab in the document area, and it vanishes in a puff of smoke.

When to Use Tabs, Tables, and Columns

Tabs can be great for laying out text in columns, but there are times when using a table or a column layout will save you time and effort.

Once you know that Pages offers newspaper-style columns, it's usually pretty easy to decide when to use them: whenever you need text that flows down one column, continues at the top of the next column, and then flows down that column in turn.

When to use a table rather than tabs can be trickier to decide. If you find yourself breaking a paragraph into multiple lines and using tabs to indent each of those lines to the same point, you will usually want to use a table instead.

Each table cell can contain one or more paragraphs and can wrap each line at the border of the cell. You can remove the borders from a table, so the effect is almost identical to using tabs. The table just provides an easier way to handle longer pieces of text than will fit on a single line.

Creating Tables

Tables are a great way to lay out data so it's clear and easy to read. A table consists of cells created by the intersection of one or more rows and one or more columns. A table can consist of a single cell, but usually most tables have multiple cells.

Most tables benefit from having one or more header or footer rows and a header column to explain the contents of the rows and columns.

You can use as many lines or paragraphs of text as you need in a cell. Pages automatically wraps each line in a paragraph within a cell, so you don't need to worry about breaking lines manually.

This section shows you everything you need to know about tables, from creating them either from scratch or from existing text and setting them up the way you want to using functions in them and sorting them.

Adding a table

To add a table, position the insertion point where you want the table, and then click the Table button on the toolbar. (Alternatively, choose Insert ➪ Table.) Pages inserts a table consisting of three columns and four rows, formats the top row as a heading row, and displays the table controls on the Format bar (see figure 4.3) and the Table pane of the Table Inspector.

The table controls on the Format bar let you do the following:

- **Align cell contents horizontally.** You can align left, center, right, and justified as usual, but you can also align automatically based on the cell content's type (for example, aligning text left but aligning numbers right).

- **Align cell contents vertically.** You can align content with the top, middle, or bottom of the cell.

- **Change the number of rows and columns.** This is a quick way to change the table's size and shape.

- **Change the fill color of selected cells.** This is a quick way to change the table's look.

Autoalign Table Cell Based on Content Type button

Justify button

Align Right button   Align Top button

Center button   Align Middle button

Align Left button   Align Bottom button

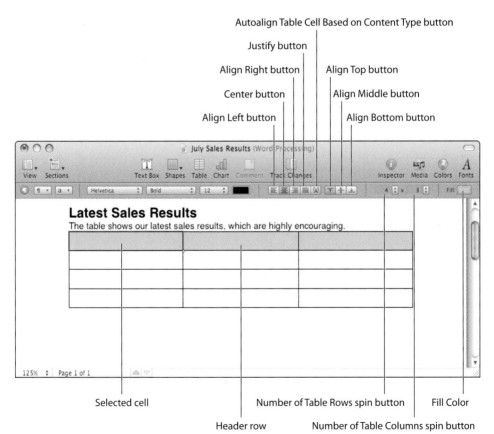

Selected cell   Number of Table Rows spin button   Fill Color

Header row   Number of Table Columns spin button

4.3 Pages inserts a standard-size table, which you can then customize as needed. The table's top row is the header row and is formatted gray.

115

Changing the number of rows or columns

Once you've inserted the table, you can easily change the number of columns and rows to suit your needs.

Until you add content to a table, the fastest and easiest way to change it is to use the Number of Table Rows spin button or the Number of Table Columns spin button. Increase the number, and Pages adds a row or column to the end of the table; decrease the number, and Pages removes the last row or column.

Genius If you have the Table Inspector open, you can also add or delete rows or columns by increasing or decreasing the number in the Body Rows box or the Body Columns box in the Table pane.

Once you've added content to a table, you can still add rows and columns this way, but Pages prevents you from using the Number of Table Rows spin button or the Number of Table Columns spin button to remove a row or column if doing so will lose the contents of one or more cells. This is a handy protective mechanism, but it means that you need to use other means to delete rows and columns that have contents.

Here are the other ways you can add and delete rows and columns:

- **Add a row.** Ctrl+click or right-click the row next to where you want to add the new row, and then choose Add Row Above or Add Row Below from the shortcut menu. You can also click in the row and then choose Format ⇨ Table ⇨ Add Row Above or Format ⇨ Table ⇨ Add Row Below.

- **Add a column.** Ctrl+click or right-click the column next to where you want to add one, and then choose Add Column Before or Add Column After from the shortcut menu. You can also click in the column, open the Format ⇨ Table submenu, and then choose the appropriate command.

Genius To quickly add a column before the current column, press Option+Left Arrow. To quickly add a column after the current column, press Option+Right Arrow.

- **Delete a column or row.** Ctrl+click or right-click anywhere in the column or row, and then choose Delete Column or Delete Row from the shortcut menu. You can also click in the column or row, open the Format ⇨ Table submenu, and then click the command there.

Resizing and rearranging a table

Before you resize a table, first click the table to display the selection handles around it. What you see will vary depending on what type of document the table is in.

- **Inline table.** If the table is inline (in a word-processing document), you'll see black selection handles on the table's bottom and right side and blue selection handles on the top and the left side. The black handles are *active*, and you can click and drag them; the blue handles are *inactive*, and won't work.

- **Floating table.** If the table is floating (in a page-layout document), you'll see black selection handles all around the table. All these handles are active, and you can click and drag any of them.

Next, to resize the table, click and drag the handles. This enables you to do the following:

- **Resize the whole table.** Click and drag a corner handle. Shift+click and drag to resize the table proportionally. Option+click and drag to resize the table about its center point. Or Option+Shift+click and drag to resize the table proportionally about its center point.

- **Change the table's width.** Click and drag a side handle to make the table wider or narrower.

- **Change the table's height.** Click and drag a handle at the bottom or top to make the table taller or shorter.

To change the width of a column, click and drag its right border. Similarly, to change the height of a row, click and drag its lower border.

You can also resize a column or row more precisely by using the Column Width box or the Row Height box in the Table pane of the Table Inspector. To resize a table itself precisely, click the table, open the Metrics Inspector, and then set the measurements you want in the Width box and the Height box.

Setting up the headers and footer for the table

When you insert a table, Pages automatically gives it a header row. You may need a header column as well or a footer row instead — or more than one column or row of headers or footers.

To change the headers for a table, use the Headers & Footer pop-up menus in the Table pane of the Table Inspector (see figure 4.4). Open the Header Columns pop-up menu, the Header Rows pop-up menu, or the Footer Rows pop-up menu and choose the number of headers you want: 0 to remove an existing header, or from 1 to 5 to have that many. In the Header Rows pop-up menu, click the Repeat Header Rows on Each Page item (putting a check mark next to it) if you want the header rows to appear on each page (if the table runs to more than one page). Repeating the header rows is usually helpful because readers don't have to refer to the previous page to see what a column represents.

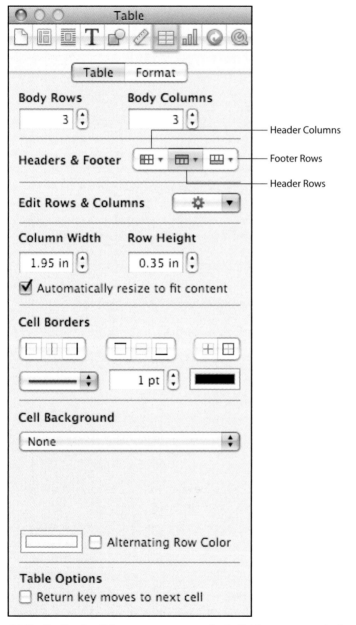

4.4 Use the Headers & Footer pop-up menus in the Table pane to set the headers and footer for a table and to control whether header rows repeat on subsequent pages.

Note You can also adjust the headers for a table by choosing Format ➪ Table and using the Header Rows submenu, the Header Columns submenu, or the Footer Rows submenu.

Converting text to a table

If the text you want to turn into a table is already in your Pages document, you can simply convert it to a table. To do so, follow these steps:

1. **Lay the text out with tabs to indicate the number of columns you want.** For example, if you use two tabs to create three columns (text, tab, text, tab, text), you'll get a three-column table.

2. **Select the paragraphs of text you want to convert.**

3. **Choose Format ➪ Table ➪ Convert Text to Table.** Pages creates a table, turning each paragraph into a row and creating columns from the divisions marked by the tabs.

4. **Add header rows or columns, or footer rows, as needed.**

Note If the paragraphs you convert to a table contain different numbers of tabs, you'll get a table with as many columns as the paragraph that has the most tabs, plus one. One fix is to move any misplaced data out of each extra column and delete it, but it's usually quicker and easier to choose Edit ➪ Undo Convert Text to Table, delete the extra tab, and then convert the text to a table again. Click the View pop-up menu button on the toolbar and choose Show Invisibles if you're having trouble finding the extra tab.

Creating a new table from an existing table

Sometimes when you've created a table, you may want to create another table from it. You can do this two ways:

- **Copy and paste the entire table.** Use this technique when you want to reuse all or most of the table. Copy the table by clicking to select it, then Ctrl+click or right-click and choose Copy from the shortcut menu. Then Ctrl+click or right-click where you want to place the copy and choose Paste from the shortcut menu. Delete any parts of the table you don't want to keep.

- **Click and drag part of the table.** Use this technique when you want to reuse just part of the table. Select the cells, rows, or columns you want, click in the selection, and then drag it to where you want the new table (see figure 4.5).

Latest Sales Results

The table shows our latest sales results, which are highly encouraging.

Year	2009	2010	2011
Sacramento			
Carson City			
Phoenix			
Nashville			

Year	2009	2010
Sacramento		
Carson City		
Phoenix		

4.5 You can quickly create a new table by selecting part of an existing table and then clicking and dragging it.

Inserting content in tables

Inserting content in a table is pretty straightforward. Try the following options:

- To type or paste in text, click the cell, and then type or paste.

- To edit the existing contents of a cell, click the cell, click to position the insertion point where you want to edit, and then make the changes. If you make a mistake, you can undo the edit by pressing Esc.

Genius

When you're entering short text items in a table, you can speed up text entry by setting Pages to automatically move to the next cell when you press Return. To do so, in the Table Options area of the Table pane of the Table Inspector, select Return key moves to next cell.

- To type a tab in a table, press Option+Tab rather than pressing Tab on its own. (Pressing Tab moves the selection to the next cell.)

- To turn existing text into a table, you can copy (or cut) and paste it, but it's often easier to convert the text to a table, as discussed earlier in this chapter.

- To insert a media item (such as an image) in a cell, drag it from the Media Browser to the Pages document. Pages inserts it as a floating object. With the object selected, click the Inline button on the Format bar to turn the object into an inline object, then click the object and drag it to the table cell in which you want it. Pages resizes the object automatically to fit the cell, but you can click it and drag a sizing handle to resize it further as needed.

Selecting parts of tables

To select a cell, click it. You can then move the selection to the next cell by pressing Tab or Right Arrow, to the previous cell by pressing Shift+Tab or Left Arrow, to the cell above by pressing Up Arrow, or to the cell below by pressing Down Arrow.

To select a block of cells, click and drag through them. Pages displays a blue border around the selection. You can also click the first cell in the block and then Shift+click the last cell to quickly select all the other cells in the block.

To select multiple individual cells, click the first, and then ⌘+click each of the others.

Converting a table to text

If you need to convert a table back to paragraphs of text, click anywhere in the table, and then choose Format⇨Table⇨Convert Table to Text. Pages creates a paragraph from each table row and separates the columns with tabs.

Merging and splitting cells

Regular tables — those with the same number of cells in each row or column — are easy to create and work with. But to present the right information exactly the way you need it, you'll often need to have different numbers of cells in different rows or columns. To create different numbers of cells, you can either merge existing cells together into a single cell or split an existing cell into two or more new cells.

Merging cells into a single cell

Merging a row of cells together into a single cell is useful for creating subheadings that span the entire table, but you can also merge cells together vertically to create a deeper cell.

To merge cells, select the cells, Ctrl+click or right-click, and then choose Merge Cells from the shortcut menu. (If you prefer, you can also choose Format⇨Table⇨Merge Cells.) Pages puts the contents of the merged cells together. If you've merged the cells horizontally, Pages divides the contents of one cell from the contents of the next using a tab. If you've merged the cells vertically, Pages divides their contents with a paragraph mark.

Splitting a cell

To split a cell into two columns or rows, click the cell, Ctrl+click or right-click it, and then choose Split into Columns or Split into Rows. (You can also choose Format⇨Table⇨Split into Columns or Format⇨Table⇨Split into Rows.)

You can then rearrange the contents of the cells as necessary by cutting and pasting them or dragging and dropping them.

Note When you split a cell into columns, Pages keeps the rest of the table as it is, simply dividing the cell you choose into two cells half the width. But when you split a cell into rows, Pages doesn't split the cell horizontally: Instead, it adds another cell of the same height below the target cell, increasing the depth of the table to accommodate it. This means that the cells you haven't split in the same row are now deeper than they were.

Formatting a table

To make your tables attractive and easy to read, you'll probably want to format them. You can apply styles and font formatting to the text as usual. Beyond that, you can change the borders, add a background, and apply a data format to the cells.

Change the table's borders

Use the controls in the Cell Borders area of the Table pane in the Table Inspector (see figure 4.6) to set up the borders that will make your table look best.

1. **In the upper row, click the button for the borders you want to affect.** Your choices are Leftmost Border, Inside Vertical Border, Rightmost Border, Topmost Border, Inside Horizontal Border, Bottommost Border, Inside Borders, or Outside Borders.

2. **Choose the style (None, Thin, Thick, Dashed, or Dotted) in the Border Style pop-up menu.**

3. **Choose the weight in points in the Border Width pop-up menu.**

4. **Choose the color by clicking the Border Color button and working in the Colors window.**

Add a background to the table

In the Cell Background area of the Table pane in the Table Inspector, choose the background type in the pop-up menu. You can then add a gradient using the techniques discussed in Chapter 1.

For more emphasis, select the Alternating Row Color check box, click the button to its left, and then choose the color to use in the background of the alternating rows.

Inside Vertical Border button

Leftmost
Border button

Topmost Border button

Inside Horizontal Border button

Border Style   Rightmost
pop-up menu   Border button   Bottommost Border button

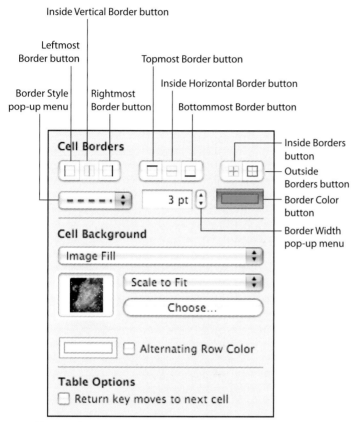

Inside Borders button

Outside Borders button

Border Color button

Border Width pop-up menu

4.6 Adding cell borders and a background can make a huge difference to a table. For extra visual impact, add an alternating row color.

Apply a cell format to the cell's contents

Apart from strictly visual formatting that controls how the contents of a cell appear, you can also apply a cell format that tells Pages the kind of information a cell contains and how to display it. For example, you can display a currency with the US dollar symbol ($), two decimal points, and the thousands separator (the comma that appears between thousands, millions, billions, and so on to make the numbers easier to read). Pages doesn't know what data you'll put in the table, so it starts each table off with the Automatic format applied to all cells. You can then use the Format pane in the Table Inspector (see figure 4.7) to apply one of the formats it offers.

123

Here are the details on the formats and how to apply them:

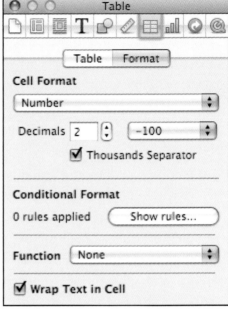

- **Automatic.** Pages applies this format when you create a table. It works well for general data, such as text. Simply choose Automatic in the Cell Format pop-up menu.

- **Number.** Use this format for general numbers, such as 100,000 or 99.05 — numbers that aren't currencies, percentages, or other specific types.

  1. Choose Number in the Cell Format pop-up menu.

  2. Set the number of decimal places in the Decimals box. For example, choose 0 to have no decimal places.

  3. In the Negative Numbers pop-up menu, choose how to represent negative numbers. Your choices are with a minus sign, in red, in parentheses, or in red parentheses.

4.7 The Format pane in the Table Inspector lets you apply cell formatting (the options differ depending on what you select in the Cell Format pop-up menu) and conditional formatting. You can also insert functions.

  4. Select the Thousands Separator check box if you want to use the thousands separator comma (for example, 1,000,000 instead of 1000000).

- **Currency.** Use this format for money values — for example, $985.16 or Can$55.00.

  1. Choose Currency in the Cell Format pop-up menu.

  2. Choose the currency symbol in the Symbol pop-up menu. For example, choose US Dollar ($).

  3. Set the number of decimal places in the Decimals box. For example, choose 2 to have two decimal places.

  4. In the Negative Numbers pop-up menu, choose how to represent negative numbers. Your choices are with a minus sign, in red, in parentheses, or in red parentheses.

5. Select the Thousands Separator check box if you want to use the thousands separator comma.

6. Select the Accounting Style check box if you want to align the currency symbol left at the beginning of the cell rather than placing it before the number.

⦿ **Percentage.** Use this format for percentages — for example, 18% or 57.9%.

1. Choose Percentage in the Cell Format pop-up menu.

2. Set the number of decimal places in the Decimals box. For example, choose 1 to have one decimal place.

3. In the Negative Numbers pop-up menu, choose how to represent negative numbers. Your choices are with a minus sign, in red, in parentheses, or in red parentheses.

4. Select the Thousands Separator check box if you want to use the thousands separator comma. Most percentages won't need it.

⦿ **Date and Time.** Use this format for dates and times.

1. Choose Date and Time in the Cell Format pop-up menu.

2. In the Date pop-up menu, choose the date format. For example, choose 01/05/2010 or Tuesday, January 5, 2010.

3. In the Time pop-up menu, choose the time format. For example, choose 20:08 or 8:08:08 PM.

⦿ **Duration.** Use this format for lengths of time, such as 10 weeks 3 days 5 hours, 1h 2m 10s 253ms, or 1:02:10.253.

1. Choose Duration in the Cell Format pop-up menu.

2. Drag the Units slider along the Wk–Day–Hr–Min–Sec–Ms axis to choose which time units to use for durations. Drag either end of the slider to add or remove units from that end. For example, select Hr–Min–Sec to measure in hours (h), minutes (m), and seconds (s), and then drag the right end of the slider to the right to add milliseconds (ms) as well.

3. In the Format pop-up menu, choose the format in which to display the duration. For example, the choices for hours, minutes, seconds, and milliseconds are 0:00:00.000, 0h 0m 0s 0ms, and 0 hours 0 minutes 0 seconds 0 milliseconds.

⦿ **Fraction.** Use this format for fractions, such as 1/2 or 677/943.

1. Choose Fraction in the Cell Format pop-up menu.

2. In the Accuracy pop-up menu, choose the accuracy. Your choices are Up to one digit, Up to two digits, Up to three digits, Halves, Quarters, Eighths, Sixteenths, Tenths, or Hundredths.

- **Numeral System.** Use this format for numbers that use bases other than ten — for example, binary (base 2) or hexadecimal (base 16).

  1. Choose Numeral System in the Cell Format pop-up menu.

  2. Choose the base number in the Base box. For example, choose 2 for binary.

  3. Choose the number of places in the Places box.

  4. For binary, octal, and hexadecimal, choose how to represent negative numbers. Select the Minus Sign option button for a conventional minus sign. Select the Two's-Complement option button to use the two's complement system of encoding negative numbers so that addition still works.

- **Scientific.** Use this format for numbers raised to a power — for example, 9.28E+08.

  1. Choose Scientific in the Cell Format pop-up menu.

  2. In the Decimals box, choose the number of decimal places. For example, choose 2 to use two decimal places.

- **Text.** Use this format for telling Pages to treat the cell's contents as text, even if it appears to be a number. Simply choose Text in the Cell Format pop-up menu; there are no further options.

- **Custom.** Use this format when you need to create custom formats for numbers and text and dates and times — for example, AD 2010-Jan-Week 2-Day 1 (Monday).

  1. Choose Custom in the Cell Format pop-up menu. Pages displays the Type text and drag elements to create a cell format dialog box (shown in figure 4.8 with options chosen for a date & time format).

  2. In the Name box, type the name you want to give the format.

  3. In the Type pop-up menu, choose the format type. Your choices are Number & Text or Date & Time.

  4. Assemble the format in the box across the middle of the dialog box. You can drag elements to where you need them, or select an element and press Delete to delete it. To add text, click in the box and type it.

  5. When you've finished, click OK to close the dialog box. You can then use your format from the Cell Format pop-up menu.

Type text and drag elements to create a cell format:

Name: AD Date

Type: Date & Time

AD 2009–Jan–Week 2–Day 1 (Monday)

AD   2009 – Jan –Week 2 –Day 1 (Monday   )

Date & Time Elements

Day of Week	Monday	Day of Year	5
Month	January	Week of Year	2
Day of Month	5	Week of Month	2
Year	2009	Day of Week in Month	1
Era	AD		
Hour	7	AM/PM	PM
Minute	08	Milliseconds	000
Second	09		

(?) (Manage Formats)   (Cancel) (OK)

4.8 The Type text and drag elements to create a cell format dialog box lets you create custom formats for numbers and text or dates and times.

Monitoring table cells for unusual values

This is a great feature. You can set Pages to monitor particular cells for particular values and apply certain formatting to any that match the criteria. *Conditional formatting* lets you do anything from automatically formatting negative values in red to getting a heads-up when a cell contains an unexpected — and perhaps incorrect — value.

To use conditional formatting, follow these steps:

1. **Select the cell or cells you want to affect.** You may want to select an entire row or column if its cells contain the same type of data. Pages applies the conditional formatting when any of the cells contains a matching value — they don't all have to match.

127

2. **Open the Format pane of the Table Inspector if it's not already open.** Click the Inspector button on the toolbar, click the Table button in the Inspector window, and then click the Format tab.

3. **In the Conditional Format area, click the Show Rules button.** Pages displays the Conditional Format dialog box. At first, this dialog box appears mostly empty, as shown in figure 4.9.

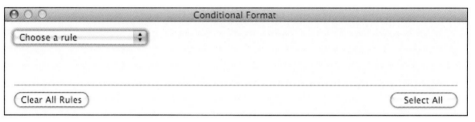

4.9 Start applying conditional formatting by clicking on the Choose a rule pop-up menu.

4. **Click on the Choose a rule pop-up menu and select the criterion for the first (or only) rule.** Your choices are: Equal to, Not equal to, Greater than, Less than, Greater than or equal to, Less than or equal to, Between, Not between, Text contains, Text doesn't contain, Text starts with, Text ends with, Text is, and With dates. This example uses the Greater than criterion.

5. **Enter the value or values in the field or fields that the Conditional Format dialog box displays.** Either simply type a value in the field, or click the blue button at the right end of the field to tell Pages you're using a cell reference, and then click the cell in the table.

6. **Change the formatting that the condition will apply.** The Sample cell in the Conditional Format dialog box shows how the cell will appear when the cell's contents meets the condition. To change it, click the Edit button and use the line of buttons that the Conditional Format dialog box displays to set the formatting (see figure 4.10). For example, click the Text color box and select a color, or click the Bold button to apply boldface. Click Done when you've finished.

7. **If necessary, click the + button at the right end of the first rule to add another rule.** Create the rule and specify formatting for it using the techniques explained in Steps 4 through 6.

8. **When you've finished setting up conditional formatting, click the red button in the upper-left corner to close the Conditional Format dialog box.**

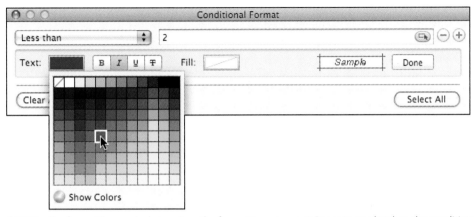

4.10 Use the formatting controls to set up the formatting you want Pages to apply when the condition is met.

When you need to remove conditional formatting from one or more cells, open the Conditional Format dialog box again to do the following:

- To see which cells have particular conditional formatting applied, click the Select All button.

- To remove a single rule, click the – button at its right end.

- To remove all rules, click the Clear All Rules button.

Genius

It can take some time and effort to get conditional formatting right, but you can save time by copying conditional formatting you've created for one table to another table.

Using functions in tables

If you want your tables to make mathematical calculations for you, insert functions in the cells that need them. Simply position the insertion point in the appropriate cell, click the Function pop-up menu in the Format pane of the Table Inspector, and choose the function you want (see figure 4.11).

Pages inserts the function to work with the numeric values in the row or column the cell is in. If Pages chooses the wrong cells, click the formula to display the Formula Editor (see figure 4.12). You can then make the following changes to the formula:

- **Change a cell reference to relative or absolute.** Click the blue oval (for example, the Employees oval in the figure), click the disclosure triangle, and then choose the reference type from the pop-up menu — for example, absolute row and column. (See Chapter 6 for the lowdown on absolute and relative references in tables.)

- **Remove an item.** Click the blue oval, and then press Delete.

- **Add cells.** Click a cell to add it to the formula, or drag through a range of cells to add them.

- **Stop editing.** When you've finished, click the green check mark to close the Formula Editor. If you've messed up and want out, click the red X instead.

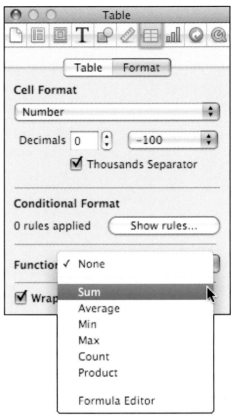

4.11 You can insert functions in seconds from the Function pop-up menu in the Format pane of the Table Inspector.

4.12 After inserting a function, you can edit the formula to make it refer to different cells.

The Function pop-up menu in the Format pane of the Table Inspector provides the most widely used formulas, but you can also type other formulas manually. See Chapter 7 for a full discussion of iWork's formulas.

Sorting a table

One of the great things about tables is that you can easily sort their rows quickly by any of the columns like this:

1. **Click in the column by which you want to sort.**

2. **Display the Table pane of the Table Inspector.** Click the Inspector button on the toolbar, click the Table button, and then click the Table tab.

3. **Click on the Edit Rows & Columns pop-up menu and choose Sort Ascending or Sort Descending.** Ascending sorts numbers from small to large, letters from A to Z, and so on. Descending sorts in reverse order: numbers from large to small, letters from Z to A, and so forth.

Hiding Rows or Columns

Sometimes it's useful to have information available in a table but hidden from view. For example, you may create a table that contains details you don't want to delete but that you don't want to share with your audience.

Pages doesn't directly let you hide rows or columns in tables. Here's the workaround:

1. Copy the table from Pages and paste it into Numbers.

2. In Numbers, hide the rows or columns you don't want to see. For example, highlight a row header, click the disclosure triangle that appears, and then choose Hide Row.

3. Copy the table from Numbers and paste it into Pages.

This is clumsy, but it's workable when you really don't want to remove the rows from the table.

Using Pages tables in Numbers or Keynote

When you've created a table in Pages, you can quickly transfer it to Numbers or Keynote to make further use of it. Simply copy the table in Pages, switch to Numbers or Keynote, and then paste the table where you want it.

Numbers or Keynote receives the entire table's data, and you can work with it as normal in that application. The one thing you lose is any comments attached to cells in the table in Pages; Numbers or Keynote doesn't receive these.

Creating Multicolumn Layouts

Newspaper-style columns of text are great for documents such as newsletters. When you reach the bottom of the first column, Pages flows the text to the top of the second column, and so on.

Pages makes it simple to use different numbers of columns in different parts of your documents. You can use multiple columns either within the body of a word-processing document or within a text box in either a word-processing document or a page-layout document.

Genius

The key to using columns successfully is to grasp what Pages calls "layouts." A *layout* is a section of a document that has particular layout characteristics — for example, two columns and one-inch layout margins. The whole of a layout must have the same number of columns, so when you need to change the number of columns, you must create a new layout.

Setting up multiple columns

Here's how to set up multiple columns.

1. **Choose the part of the document you want to affect.**

   - To affect the whole document, just click anywhere in it.

   - To affect just part of a document, cordon it off with layout breaks. Click before the first paragraph and choose Insert ⇨ Layout Break to insert a layout break. Then click after the last paragraph and use the command again. Click in the part you just cordoned off.

- To change the number of columns from the point you choose forward, without affecting the text earlier in the document, insert a layout break by choosing Insert ⇨ Layout Break.

2. **Open the Layout pane in the Layout Inspector (see figure 4.13).** Click the Inspector button on the toolbar or press ⌘+Option+I to open the Inspector window. Click the Layout button to display the Layout Inspector. Make sure the Layout pane is displayed. (If it's not, click the Layout tab.)

3. **In the Columns box, set the number of columns you want.** Usually, it's easiest to click the spin buttons to increase or decrease the number of columns, but you can type the number if you prefer.

4. **Select the Equal column width check box if you want to make each column the same width.** If you want columns of different widths, deselect this check box, and then change the column width in one of two ways:

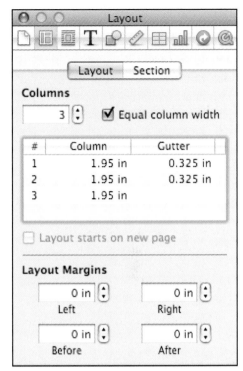

4.13 Use the Layout pane of the Layout Inspector to set the number of columns, change their width, and adjust the width of the gutter between them.

- In the Layout pane, double-click the Column box you want to change, and then type the width.

- In the document, click in the column you want to change, and then drag the margin marks. This method is easier once you have entered text in the columns in the document.

5. **If you need to adjust the distance between columns, double-click the Gutter measurement for a column, and then type the new value.** If you've selected the Equal column width check box, Pages automatically makes each of the gutter measurements the same.

6. **If you need to make the layout start on a new page rather than on the current page, select the Layout starts on new page check box.** This setting is handy when you need to create a new page for the layout, but you won't need it every time.

7. **If you want to use different margins around this layout, set them in the Left box, Right box, Before box, and After box in the Layout Margins area.** For example, you may want to give this section extra indentation from the left and right, or extra space above and below.

8. **When you're satisfied with the columns, close the Layout Inspector.**

Now that you've set up the columns, enter the text in them. When the text reaches the bottom of one column, it flows automatically to the next column.

Inserting column breaks and layout breaks

If you need to end a column before the bottom of a layout (for example, so that a particular paragraph appears at the top of the next column), insert a column break. Position the insertion point and choose Insert ⇨ Column Break.

To create a new layout in a document, insert a layout break at the point where you want the new layout to begin. Just place the insertion point and choose Insert ⇨ Layout Break.

Layout breaks and column breaks are invisible characters, so you don't see them unless you've set Pages to display invisibles. (Click the View pop-up menu button on the toolbar and choose Show Invisibles or press ⌘+Shift+I.) When you do display invisibles, these breaks appear as horizontal lines with the boxes, as shown in figure 4.14.

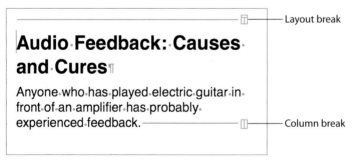

Layout break

Audio·Feedback:·Causes· and·Cures¶

Anyone·who·has·played·electric·guitar·in· front·of·an·amplifier·has·probably· experienced·feedback.———————— Column break

4.14 Display invisible characters when you need to see exactly where layout breaks and column breaks are.

To delete a column break or a layout break, position the insertion point before it and then press Delete.

Adding Images, Shapes, and Charts

To spruce up your Pages documents, you can add images, shapes, and charts by using the techniques described in Chapter 1. This section covers extra options you have and considerations to take in to account when using these features in Pages.

Choosing between inline objects and floating objects

Pages lets you insert an image or other object as either an inline object or a floating object:

- An *inline object* is anchored at a particular point in the text, so that it acts like one of the characters in the text. If you delete characters before the inline object, it moves toward the beginning of the document. If you insert characters before the inline object, it moves toward the end of the document.

- A *floating object* is anchored to the position you choose on the page and remains there unless you move or delete it. A floating object isn't part of the text of the document.

Choose an inline object when you want the object to appear in or near to a particular paragraph. For example, when you're using a photo to illustrate a story in a newsletter, make the photo an inline object so that it appears at the right point in the text.

Choose a floating object when you need to make sure that it appears in the right position in the layout, even if the text moves up or down the document.

If you realize you've chosen the wrong type of object, you can quickly convert it to the other type. Simply click the object, and then click the Floating button or the Inline button on the Format bar, as needed. Alternatively, open the Inspector window, click the Wrap Inspector button to display the Wrap Inspector (see figure 4.15), and then select either the Inline (moves with text) option button or the Floating (doesn't move with text) option button.

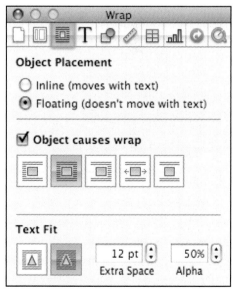

4.15 You can use the Wrap Inspector to switch an image or other object from inline placement to floating placement.

When you switch a floating object to an inline object, the application may need to move the object to position it in the text. When you switch an inline object to a floating object, it stays where it is unless you move it.

Genius

You can quickly select all floating objects by clicking one floating object and then pressing ⌘+A. When you've finished working with the floating objects, press ⌘+Shift+A to deselect them.

Wrapping text around an image or object

When you've placed an image or object in a document, you need to decide how to wrap text around it.

The quick way to set wrapping is to click the object, click the Text Wrapping pop-up menu button at the right end of the Format bar, and choose the wrapping you want. (When you select an object, the Format bar displays controls for formatting the object.) Figure 4.16 shows (from top to bottom) the options on this menu, where you can choose to wrap text left, around, right, above or below (where there is the most space), above and below, or to remove the wrap.

4.16 You can quickly set text wrapping from the Format bar.

To adjust wrapping further, follow these steps:

1. **Click the image or object, and then click the Inspector button on the toolbar to open the Inspector window.**

2. **Click the Wrap Inspector button on the toolbar to display the Wrap Inspector.**

3. **If necessary, change the object's placement by clicking the Inline option button or the Floating option button.** Usually, you'll have chosen this setting before displaying the Wrap Inspector.

4. **Select the Object causes wrap check box and then click the button for the wrapping you want.**

5. **In the Text Fit area, as shown in figure 4.17, choose between the Wrap Text to Follow a Square Border button and the Wrap Text to Follow the Object's Contours button.**

6. **To change the amount of space between the object and the text, increase or decrease the number of points in the Extra Space box.**

7. **If you're using the Wrap Text to Follow the Object's Contours setting and you want to make more or less of the object's background transparent, increase or decrease the percentage in the Alpha box.** (See Chapter 1 for more details on making backgrounds transparent.)

8. **Close the Inspector window when you've finished.**

Adding charts

You can insert a chart in a Pages document by using the techniques explained in Chapter 1. There are two main points to keep in mind:

⬤ **Pages automatically formats the chart to suit the template you're using.** Check whether this formatting suits the look of the document you're creating. If not, simply use the controls in the Chart Inspector to override the automatic formatting with a look that you prefer.

⬤ **You may need to change the chart's wrapping.** If you insert a chart in a word-processing document as an inline object, you can change it to a floating object if you want to wrap it differently. To change the wrapping, follow the instructions in the previous section.

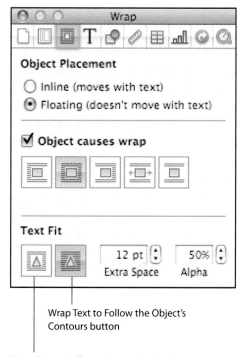

Wrap Text to Follow the Object's Contours button

Wrap Text to Follow a Square Border button

4.17 The Wrap Inspector provides close control over text wrapping.

Flowing Text through Linked Text Boxes

When you need to place text in precise areas of a document, create text boxes and enter text in them. You can link two or more text boxes into a series so that text flows from the first text box to the second (and so on) without your having to adjust the amount of text in any text box.

Text boxes tend to be more useful in page-layout documents than in word-processing documents, but you can use them in word-processing documents as well if you need them.

Placing text boxes in a document

Start by placing the first text box in the document like this:

1. **Click the Text Box button on the toolbar or choose Insert ➪ Text Box.** Pages adds a standard-size text box to the middle of the document (see figure 4.18).

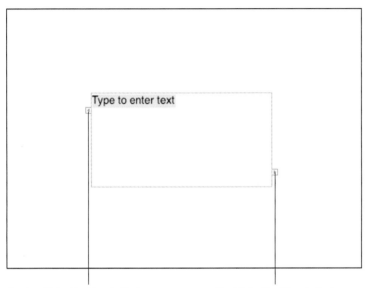

Previous linked text box indicator   Next linked text box indicator

4.18 Pages inserts a standard-size text box in the middle of the document. The blue indicators are not fully filled in, meaning that the text box has no link to a previous text box or next text box.

Genius

If you find you've placed your linked text boxes in the wrong sequence, either simply reverse their positions, or delete the text box you don't want. Any text that was in that text box flows back to the previous text box, so you don't lose it. (If you delete the last remaining text box, you do delete the text.) You can then insert a replacement text box where you need it.

2. **Resize the text box if necessary.** Click the text box to display its sizing handles, and then click and drag a handle to change the size. You can use iWork's standard tricks:

- Shift+drag to resize the text box proportionally.

- Option+drag to resize the text box about its center.

- Shift+Option+drag to resize the text box proportionally about its center.

- ⌘+drag to rotate the text box.

- ⌘+Option+drag to rotate the text box around the opposite handle.

- ⌘+Shift+drag to rotate the text box in 45-degree increments.

Note From the keyboard, press ⌘+Return to stop editing a text box and select its frame.

3. **Add the text to the text box.** You can type it, paste it with its formatting, or use the Paste and Match Style command to make the text match the style of the text box.

Genius When you're working with linked text boxes, it's often easier to write and edit the text in another Pages document where you can see all the text at once. When it's finished, copy the text to the text boxes. You may then need to edit it for length, but that won't take long.

Flowing text between text boxes

If you've entered too much text to fit in the text box, Pages displays a clipping indicator at the bottom of the text box (see figure 4.19).

To make text flow to another text box, you simply link this text box to another text box. You can either create a new text box or use an existing text box. Once you've linked the text boxes, you can work with their contents as a single stream. You can also break the links if you no longer need them.

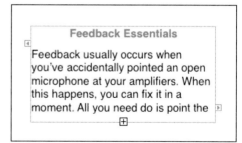

4.19 The clipping indicator tells you that there's too much text to fit in the text box.

Creating a link to another text box

Here's how to link to another text box:

1. **Select the text box you've been working with.** For example, click the text box, or press ⌘+Return if you've been working in it.

2. **Choose whether to link to a text box before this one or after this one.** Click the Next linked text box indicator if you want the content in the text box you're linking to be after the content in this text box. If you want the content in the text box you're linking to be before the content in this text box, click the Previous linked text box indicator. Pages displays a + sign next to the mouse pointer and a pop-up instruction message (see figure 4.20).

3. **Choose whether to create a new text box or link to an existing one:**

   - **Create a new text box.** Click where you want to place the new text box. Pages adds the text box, flows the text, and shows the link with a blue line (see figure 4.21).

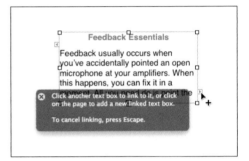

4.20 You can link a text box either to a new text box or to an existing text box.

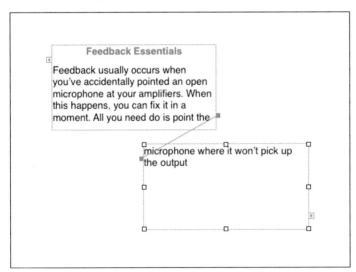

4.21 Pages shows you clearly which text box is linked to which.

- **Link to an existing text box.** Move the mouse pointer over the text box so that Pages displays a blue border around it, and then click. Pages links the text box, flows the text, and displays a blue line showing the link.

Working with text in linked text boxes

After you link the text boxes, Pages treats their combined text as a single, continuous stream. So when you add text to the first linked text box, any text that no longer fits into it flows down into the second text box, and so on down the chain of links. Similarly, if you delete text from the second text box, Pages flows text back from the third and subsequent text boxes so that the second text box remains filled.

You can format the text in a text box using Pages' regular formatting features. For example, apply the main formatting using paragraph styles; add character styles to text that requires emphasis; and then apply direct formatting to any text that needs special treatment.

To adjust the distance between the border of the text box and the beginning of the text, open the Text Inspector and drag the Inset Margin slider in the Text pane. For finer adjustments, click the spin buttons on the box in which the measurement appears.

Breaking the connection between two text boxes

To break the connection between two text boxes, click the Next linked text box indicator or the Previous linked text box indicator and drag it off the text box into the document. The connection disappears in Mac OS X's trademark puff of smoke, and the text disappears from the text box, flowing into the previous text box or next text box, depending on which connection you deleted.

Genius

Breaking the connection to a linked text box disconnects any text boxes farther down the chain. For example, if you have a chain of three linked text boxes, disconnecting the second from the first disconnects the third as well.

Cutting a linked text box out of the chain

Instead of disconnecting text boxes, you may sometimes need to cut a linked text box out of the chain. For example, you may need to replace a text box with an image to make your page layout work effectively.

To cut a linked text box out of the chain, simply click it and press Delete. Pages removes the text box and reflows its content through the other text boxes in the chain. For example, if you have a chain of four text boxes and delete the second one, the third text box becomes the second text box and receives the text that was previously in the second one, and so on down the chain.

Now That I've Made My Document, How Can I Use It?

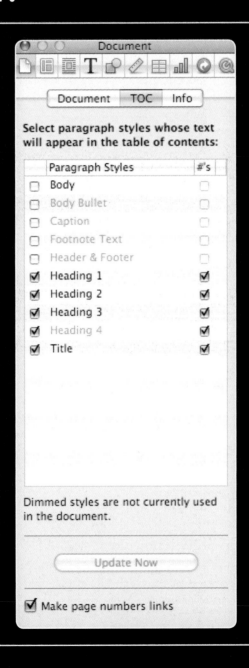

After creating a document, you're probably ready to share it with others. First, work with your colleagues to finalize the document, using Pages' powerful features for commenting on documents and tracking changes, reviewing the changes, and creating a final version. Next, you can add polish to the document by inserting footnotes or endnotes, or a table of contents; by linking different parts of the document for easy navigation; and by controlling hyphenation in the document. You may also want to use Pages' automated Proofreader feature — preferably with care. Once you've done that, you'll be ready to share the document either on- or offline using a variety of formats.

Using Track Changes

When you're editing a document on your own, you can usually make changes freely without needing to show what they are. But when you're editing a document with other people, it's often helpful to track your changes so anyone else editing the document can review them. Or, if you need to review changes your colleagues have made to a document, those changes are apparent.

Comparing an edited version of a document to the original is slow, laborious work, even for a seasoned editor. Pages' Track Changes feature makes it easy for you to see exactly which changes your colleagues have made to a document.

Turning on Track Changes

To turn on Track Changes, choose Edit ➪ Track Changes.

Genius

If you work extensively with Track Changes, add the Track Changes button to the toolbar so that you have instant access to it. Choose View ➪ Customize Toolbar, drag the Track Changes button to a convenient spot on the toolbar, and then click Done.

Pages displays the tracking bar below the Format Bar (if it's displayed) and also displays the Comments and Changes pane on the left side of the window (see figure 5.1). The tracking bar contains the following controls:

- **Show/Hide Comments and Changes pane button.** Click this button to hide the Comments and Changes pane or show it again.

- **Tracking Bubbles pop-up menu.** Choose whether to show bubbles for the tracked changes and, if so, which ones.

- **Tracking switch.** Move this switch to Paused to turn off tracking temporarily.

- **Previous Change button.** Click this button to move to the previous tracked change.

- **Next Change button.** Click this button to move to the next tracked change.

- **Accept button.** Click this button to accept the selected change.

- **Reject button.** Click this button to reject the selected change.

- **Markup View pop-up menu.** Use this menu to switch among viewing the markup, viewing the markup without deletions, and viewing the final text without the markup or deletions.

- **Change Tracking Action pop-up menu.** Use this menu to take other actions, such as turning off tracking or changing the color used for an author's changes.

Show/Hide Comments and Changes pane

Tracking Bubbles pop-up menu   Accept button

Next Change button   Reject button

Previous Change button   Tracking bar

Tracking Switch   Markup View pop-up menu

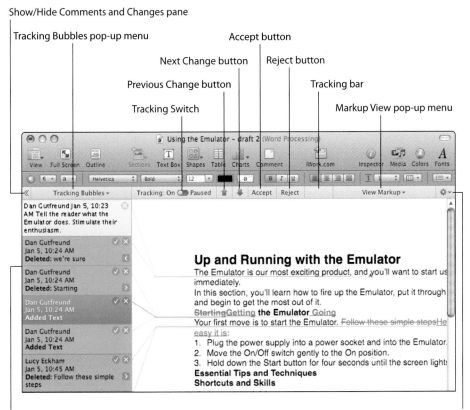

Comments and Changes pane   Change-Tracking Action pop-up menu

5.1 The tracking bar appears below the Format Bar and gives you one-click control of Track Changes. The Comments and Changes pane shows the details of the changes.

Choosing which marks and color to use

To choose which marks and colors to use for inserted text and deleted text, open Pages' preferences (press ⌘+, [comma] or choose Pages➪Preferences), click the General button to display the General pane, and choose settings in the Deleted Text pop-up menu and the Inserted Text pop-up menu. Double underline for inserted text and strikethrough for deleted text are usually the easiest choices for reading the document. Pages' default setting for inserted text is single underline.

Pages automatically assigns a different color to the first seven people who edit a document with Track Changes on. To change the color used for the changes you're making, click the Markup View button at the right end of the tracking bar, highlight Select Author Color, and then click the color you want to use.

Genius

Sometimes you may need to set up Pages to track some changes in a different color so that you can identify them easily. To do this, save and close the document, change the name shown in the Author box in Pages' preferences, and then reopen the document. Pages then tracks the changes using the new author name and a different color. If you simply change the author name without closing the document, Pages also assigns the changes you had already made using your current author name to the new author name you enter in the General pane in the Preferences window.

Making untracked changes while Track Changes is on

Even after you've turned Track Changes on, you'll often need to make changes without having them tracked. For example, using Replace to replace every instance of a widely used phrase in a document with Track Changes on produces a slew of insertions and deletions. If these will be distracting, you can pause tracking while you make the changes.

To pause tracking, click the Tracking switch on the tracking bar to move it to the Paused position. (You don't have to click the switch — clicking anywhere on the button does the trick.) Make the changes; then, when you're ready to resume tracking, click the button again to move the switch back to the On position.

Caution

If you're used to Microsoft Word, you'll notice that pausing Track Changes is a bit different. In Word, you can turn Track Changes on and off freely with the click of a mouse. In Pages, you can only pause Track Changes; when you want to turn Track Changes off fully, you must immediately accept or reject all the tracked changes.

Understanding which changes are tracked and which aren't

After you turn on Track Changes for a document, Pages tracks text you insert and text you delete. For example, if you type a new word, Pages marks it as an insertion (using the formatting you chose in the Inserted Text pop-up menu in Pages' preferences). And if you select a sentence and then press Delete, Pages marks it as a deletion (using the formatting you chose in the Deleted Text pop-up menu).

Pages tracks almost all the changes you make to the formatting of a document. For example, if you apply a different style or italic formatting, Pages tracks the change. However, Pages doesn't track the changes you make to styles themselves. For example, if you modify a paragraph style or create a new character style, Pages doesn't track the change.

Note When you move text by cutting it from one point in a document and pasting it into another point, Pages marks the changes as a deletion (where you cut) and an insertion (where you pasted). Other word-processing applications (for example, Microsoft Word) mark a move operation within the same document differently, enabling you to see that some text has come from within the document.

Pages '09 also tracks changes to inline objects and floating objects in your documents. This is a big improvement over Pages '08, which tracks only changes to inline objects, not to floating objects. For example, if you add a floating text box, Pages '09 marks it as Added: Text Box, whereas Pages '08 doesn't mark the added text box but does mark new text that you insert in it.

Exchanging tracked changes with Microsoft Word

Pages' Track Changes feature is compatible with Microsoft Word's Track Changes feature most of the time, so you can export a Pages document to Word format with the tracked changes in place. (See the section on sharing your documents, later in this chapter, for instructions on saving a Pages document as a Word document.) Similarly, you can open a Microsoft Word document in Pages and see the tracked changes and comments it contains.

Caution If a Word document contains very many tracked changes, Pages may crash while opening the document. If you know that a Word document you're about to open contains extensive tracked changes, save and close any Pages documents you're working on before you open it.

If you're receiving Word documents with tracked changes from a Windows user, make sure the documents are saved in the Word 97–2003 document format (with the .doc file extension) rather than in the Word 2007 document format (with the .docx file extension). The problem with the Word 2007 document format is that it includes two types of change tracking features that Pages doesn't support:

- **Moved text.** Word 2007 lets you mark text moved within the document (using Moved From and Moved To marking) rather than marking it as a deletion and an insertion. Pages loses both the Moved From text and the Moved To text, which can give you severe headaches.

- **Table cell highlighting.** Word 2007 lets you mark inserted, deleted, merged, and split cells in tables. Pages can't handle these tracked changes.

The easiest way to solve this problem is to open the Word 2007 document in Word and use the File ⇨ Save As command to save it as a Word 97–2003 document. When you do this, Word automatically converts the moved-text markings to insertions and deletions and removes table cell highlighting that the Word 97–2003 document format does not support.

Note Although Word 2008 for the Mac also uses the .docx document format, it doesn't let you track moved text or table cell highlighting. So Word 2008 documents in .docx format don't cause the same problems for Pages that Word 2007 documents in .docx format cause.

Hiding and showing tracked changes

When you're working with a document that contains extensive tracked changes, it's often helpful to hide the tracked changes so that you can focus on the document's current text.

If you want to hide all the markup, click on the Markup View pop-up menu at the right end of the tracking bar and choose View Final. When you want to see the markup again, click on the Markup View pop-up menu and choose View Markup. The menu's button on the tracking bar shows the name of the current view, so you can easily tell what you're looking at.

If you just want to hide the deleted text, click on the Markup View pop-up menu and choose View Markup Without Deletions. When you want to see the deleted text again, click on the Markup View pop-up menu and choose View Markup.

Similarly, you can hide all change bubbles or hide some of them by clicking on the Tracking Bubbles pop-up menu at the left end of the tracking bar. You can choose one of these three options to control the display of change bubbles overall:

- **Show All.** This setting lets you see every change bubble in the document and get an overview of the "damage" done.
- **Show Only for Selection.** This setting is great for focusing on a particular paragraph. It's especially useful when the document has been heavily edited and contains lots of change bubbles.
- **Hide All.** This setting hides all the change bubbles but doesn't hide the Comments and Changes pane. Use this setting when you want to work with comments.

Whether you're using the Show All setting or the Show Only for Selection setting, you can turn off the display of formatting bubbles by clicking on the Tracking Bubbles pop-up menu and choosing Show Formatting Bubbles, thus removing the check mark from it. Repeat the command to show the formatting bubbles again (displaying the check mark on the Show Formatting Bubbles menu item).

Reviewing tracked changes

If you're the one circulating the document with tracked changes to your colleagues, you'll probably want to review those tracked changes and decide which ones to accept and which to reject.

Usually, it's easiest to do this using the Comments and Changes pane (see figure 5.2). Each change bubble contains an Accept Change button (the button with the check mark) and a Reject Change button (the button with the X). When you're using the View Markup setting or the View Final setting, the change bubble for a deletion also includes a Show or Hide Deleted Text button (the button with the arrow) that you can click to toggle the display of the deletion.

If the Comments and Changes pane is hidden, display it by clicking the Show/Hide Comments and Changes Pane button at the left end of the tracking bar. You can then simply click the Accept Change button to accept a change and integrate it into the document, or click the Reject Change button to reject a change and remove it from the document.

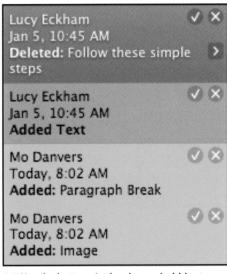

5.2 Use the buttons in the change bubbles to work quickly through the tracked changes in a document.

The other way to work through the tracked changes is to click the Next Change button or the Previous Change button on the tracking bar. Pages highlights the next change or previous change in the document, so you can see clearly what it is. You can then click the Accept button or Reject button on the tracking bar or click the Accept Change button or Reject Change button in the change bubble.

When a paragraph or other section of text contains many changes, you may want to deal with them all at once rather than one by one. Select the paragraph or section, click the Action button at the right end of the tracking bar, and then click Accept Selected Changes or Reject Selected Changes.

Other times, you may need to accept or reject all the changes in a document at once. To do so, click on the Action pop-up menu at the right end of the tracking bar, and then click Accept All Changes or Reject All Changes. Use these commands with care, as Pages doesn't make you confirm your action. If you choose either command by mistake, press ⌘+Z or choose Edit ⇨ Undo Accept All Changes or Edit ⇨ Undo Reject All Changes immediately.

Turning off Track Changes and accepting or rejecting changes

To turn off Track Changes, either choose Edit ⇨ Turn Off Tracking or click on the Action pop-up menu at the right end of the tracking bar and choose Turn Off Tracking. When you do this, Pages displays a dialog box asking whether you want to accept all the changes in the document or reject them all (see figure 5.3).

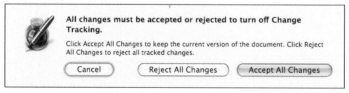

5.3 When you turn off Track Changes, you must either accept all the tracked changes in the document or reject them.

Click the appropriate button, and then save the document. If you don't want to accept or reject all the changes just yet, click Cancel, and then work your way through the changes one by one (or group by group) before turning off Track Changes.

Saving a copy of a document as final

To save a copy of a document as final, click on the Action pop-up menu at the right end of the tracking bar and choose Save a Copy as Final. In the Save As dialog box that Pages displays, type a name for the final version of the document, and then click Save.

Using Comments

Track Changes is great when you need to make changes right in the text of the document, but if you want to convey your thoughts about the document's contents or structure, or pose an argument to your colleagues, you're better off using comments.

Here's how to add a comment to a document:

1. **Click or select the item to which you want to attach the comment.** Pages calls this the *comment anchor*.

2. **Click the Comment button on the toolbar.** (You can also choose Insert ⇨ Comment from the menu bar.) Pages puts a box around the object or text you selected, opens the Comments and Changes pane on the left of the window if it was hidden, and starts a new comment beginning with your author name (as it is set in the General preferences), the date, and the time.

3. **Type the text of the comment after this information (see figure 5.4), or paste in text that you've copied from elsewhere.** You can format the comment text as needed — including using styles if you want. You can also delete the text that Pages automatically adds, but normally it's helpful to have it in the document so that the author can see who made which comments.

Mo Danvers Today, 10:54 AM ⊗
Can we come up with something snappier than this? "Up and Running" isn't catchy.

Up and Running with
The Emulator is our most exciting
using it immediately.
In this section, you'll learn how to f
and begin to get the most out of it.

5.4 Type the text of your comment after the name, date, and time that Pages inserts in the comment bubble. A line leads from the comment bubble to the comment anchor in the document.

When you need to delete a comment, click its Delete button — the round button with an X in the comment bubble.

Genius

Even though the bubbles for tracked changes appear in the Comments and Changes pane, comments are separate from tracked changes. When you accept or reject all tracked changes, comments are unaffected.

Adding Footnotes or Endnotes

If you need to provide references to the material your document cites, you can add notes to it. Pages enables you to create footnotes and endnotes containing text, images, and even tables.

Choosing between footnotes and endnotes

First, choose which type of notes you want to use:

- **Footnotes.** A footnote appears at the bottom of the page that includes the text to which the note refers. (Sometimes a footnote may need to be continued to the next page of the document.) Footnotes are great for providing reference information the reader can find easily, without having to turn pages, but if you have many footnotes, long footnotes, or both, page layout can become difficult.

- **Endnotes.** An endnote appears at the end of the document that includes the text to which the note refers. If the document is broken up into sections, you can place the endnotes at the end of each section instead. Choose endnotes when you don't want to shorten your pages with notes, when you are including notes that you think many readers will not bother to read (for example, academic citations), or when the document contains enough notes to make page layout tricky.

Unlike most other word-processing applications, Pages doesn't let you use both footnotes and endnotes in the same document. This saves confusion, because once you've started using one type of notes, you can't accidentally use the other type. However, you can convert all footnotes to endnotes, or all endnotes to footnotes, so starting with the wrong type of note isn't a disaster.

Genius

You can use footnotes and endnotes only in word-processing documents, not in page-layout documents. If you need to add notes to a page-layout document, you must do them the hard way: Add a superscript number (or a symbol) to mark the note, and then add a text box containing the note text at the bottom of the page or the end of the document.

Pages comes set to automatically use footnotes in a word-processing document, so if you want footnotes that use standard numbering (1, 2, 3, and so on, continuing through the whole document), you're all set. If you want endnotes instead, or if you want to choose a different type of numbering, follow these steps:

1. **Open the document if it's not already open.**

2. **Click the Inspector button on the toolbar to open the Inspector window. Click the Document Inspector button to display the Document Inspector, and then click the Document tab if necessary to display the Document pane (see figure 5.5).**

3. **In the Footnotes & Endnotes pop-up menu, choose the type of notes you want:**

   - **Use Footnotes.** Select this to have footnotes in the document.

   - **Use Section Endnotes.** Select this to use endnotes that appear at the end of the section to which they belong. This is useful if your document contains different chapters and you want the endnotes at the end of the chapters.

   - **Use Document Endnotes.** Select this to use endnotes that appear at the end of the document. This is the typical place to put endnotes in many types of documents.

4. **In the Format pop-up menu, choose the number format you want.** Your choices are 1, 2, 3; i, ii, iii; or symbols (the asterisk, the dagger character, the double-dagger character, the section-mark character, and so on). Symbols can be effective for footnotes but are highly confusing for endnotes because the reader must know the page number as well as the symbol.

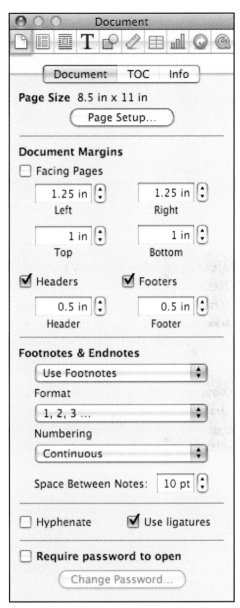

5.5 Use the controls in the Footnotes & Endnotes section of the Document pane to choose which type of notes your document has, how to number them, and whether the numbering is continuous.

Genius

If you want to check quickly which type of notes a document uses, click on the Insert menu. You'll see one of these three commands: the Footnote command, the Section Endnote command, or the Document Endnote command. The command tells you the type of note.

5. **In the Numbering pop-up menu, choose how to number the notes.** Your choices are:

- **Continuous.** Pages increases the numbers throughout the document or section. This is the best choice when your document contains many notes.

- **Restarts on Each Page.** Choose this option when you're using symbols to identify footnotes. Don't use this option for endnotes — it's a recipe for confusion.

- **Restarts for Each Section.** Choose this option when your document is divided into sections and you're placing endnotes at the end of each one rather than the end of the document.

6. **In the Space Between Notes box, you can adjust the amount of space to leave between notes.** The default value, 10 points, works well for many documents.

Close the Inspector window when you've finished making your choices.

Inserting a footnote or endnote

Inserting a footnote or endnote takes only a moment. Simply click in the text of the document where you want the note's reference mark to appear, open the Insert menu, and then choose the appropriate note command: Footnote, Section Endnote, or Document Endnote. (Only one of these commands appears on the menu, reflecting the choice you made in the Document Inspector.)

Pages adds the note's reference number or mark to the document's text and then creates a new text box in the footnote or endnote area so that you can add the note's contents. Type or paste in the text you want for the footnotes.

Genius

You can also use custom symbols for your footnotes and endnotes. Open the Insert menu, then press Option and choose Custom Footnote, Custom Section Endnote, or Custom Document Endnote. In the dialog box that appears, pick a custom mark for the note. To change the numbering on an existing footnote, right-click or Ctrl+click it, and then choose Use Custom Mark. To switch a note from a custom mark to automatic numbering, right-click or Ctrl+click the mark and choose Use Automatic Numbering.

Pages separates the footnote area or section endnote area from the main text of the document using a horizontal line. When you're using document endnotes, Pages places the notes on a new page at the end of the document.

Working with footnotes and endnotes

In many documents, footnotes and endnotes require only plain text, but Pages lets you use all kinds of formatting in your notes if you need it. You can also insert images, tables, and other objects if necessary. Bear in mind that any large object may wreak havoc with the page layout of your document; if you need to create complex notes, endnotes are often a better choice than footnotes.

You can jump to a footnote or endnote by double-clicking its reference number or symbol in the document. Similarly, you can jump to the reference number or symbol in the document by double-clicking the number or symbol in the footnotes or endnotes area.

To delete a footnote or endnote, just select its reference mark in the document, and then press Delete. Pages removes both the reference mark and the note in the footnotes or endnotes section.

Converting footnotes to endnotes — or vice versa

If you find you need to use a different type of notes in your document, you can convert your existing notes by simply choosing the kind of notes you want in the Footnotes & Endnotes pop-up menu in the Document pane of the Document Inspector. Pages converts your existing notes to the kind you choose.

Using Bookmarks to Link Parts of a Document

In many documents, it's handy to link related but separate parts together so that anyone reading the document online can quickly jump to another relevant part. You can create such a link by adding a bookmark to the part of the document you want to reach and then inserting a hyperlink to that bookmark.

Note
You can place bookmarks only in the main text of a document, not in other items such as text boxes, tables, or footnotes or endnotes.

Adding a bookmark

Here's how to add a bookmark:

1. **Select the part of the document that you want to bookmark.** For example, select a word or a sentence. If you want the bookmark to mark a single point, click at that point.

2. **Choose Insert ➪ Bookmark.** Pages opens the Link Inspector (if it's not already open) and displays the Bookmark pane (see figure 5.6).

3. **Rename the bookmark if necessary.** If you selected text, Pages automatically names the bookmark from the text. If you chose to mark a point, Pages adds a generic name such as Bookmark. Double-click the bookmark, type the name you want to give it, and then press Return to apply the name.

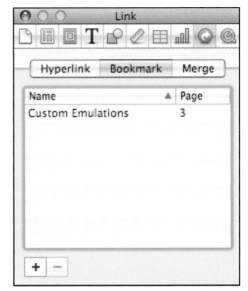

5.6 Use the Bookmark pane in the Link Inspector to insert a bookmark in a document. You can then create a hyperlink to jump to the bookmark.

Note Once you've opened the Bookmark pane of the Link Inspector, you can quickly add another bookmark by selecting the text you want to mark, and then clicking the Add (+) button to add a new bookmark to the list.

Leave the Link Inspector open for the moment so that you can add the hyperlink to the bookmark, as described next.

Note To delete a bookmark, open the Bookmark pane of the Link Inspector, click the bookmark, and then click the Delete (–) button.

Adding a hyperlink to a bookmark

Now that you've placed the bookmark, you can create a hyperlink that jumps to it. Follow these steps:

1. **In the document, navigate to where you want to insert the hyperlink.**

2. **Type the text you want to use for the hyperlink, and then select the text.**

3. **Click the Hyperlink button in the Link Inspector to display the Hyperlink pane.**

4. **Select the Enable as a hyperlink check box.** Pages turns the text you selected into a hyperlink.

5. **In the Link To pop-up menu, choose Bookmark.**

6. **In the Name pop-up menu, choose the name of the bookmark (see figure 5.7).**

You can now click the hyperlink in the document to jump straight to the bookmark.

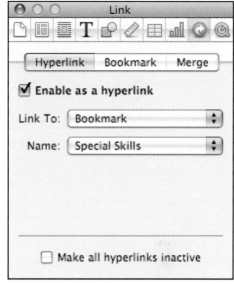

5.7 From the Hyperlink pane of the Link Inspector, you can create a link that goes to a bookmark.

Note

When you export a document to a PDF file, Pages maintains the links to bookmarks so readers can continue to use them in the PDF.

Controlling Automatic Hyphenation

To prevent awkward breaks at the ends of lines in your documents, Pages automatically inserts hyphens to break words. Normally, this automatic hyphenation is helpful, but sometimes you may want to prevent Pages from breaking words in a particular paragraph, or you may want to tell Pages to break a certain word in a way that automatic hyphenation may not get right.

To turn off automatic hyphenation, follow these steps:

1. **Select the paragraph or paragraphs you want to affect.** If you want to adjust hyphenation only for a single paragraph, click in that paragraph. If you want to affect multiple paragraphs, select them.

2. **Click the Inspector button on the toolbar to display the Inspector window (unless it's already displayed).**

3. **Click the Text Inspector button in the toolbar to display the Text Inspector.**

4. **Click the More button to display the More pane.**

5. **Near the bottom of the pane, select the Remove hyphenation for paragraph check box.**

6. **Close the Inspector window unless you need to use it further.**

Genius To prevent Pages from hyphenating a particular word, Ctrl+click or right-click the word and choose Never Hyphenate from the shortcut menu. To permit hyphenation again, Ctrl+click or right-click the word and choose Allow Hyphenation.

Adding a Table of Contents

In a long document, it's often useful to have a table of contents to show the reader a list of the different chapters or sections and the pages on which to find them. Pages uses paragraph styles to identify the paragraphs that should appear in the table of contents. You can customize the table of contents by choosing which paragraph styles to use.

Genius You can insert multiple tables of contents in a Pages document, so you can create one for each chapter or section of a document if necessary. Each table of contents contains the headings (or other paragraphs you choose to include) from the part of the document after it and up to the next table of contents. So if you want a single table of contents, place it at the start of the document, before any of the headings you want to include.

Inserting a table of contents

To insert a table of contents, simply position the insertion point where you want the table of contents, and then choose Insert ⇨ Table of Contents. Pages creates the table of contents (or TOC) and displays the TOC pane of the Document Inspector so that you can customize it by selecting the check box for each style you want to include. Dimmed styles are ones you haven't yet used in the document.

Note If Pages doesn't find any paragraphs for the table of contents, it displays a message explaining that the table of contents is empty because the document doesn't use any of the paragraph styles. If you're sure you *have* used some of the paragraph styles, the problem is most likely that the table of contents is positioned after the paragraphs in the document rather than before them.

Formatting a table of contents

Pages automatically formats the table of contents using the TOC Heading 1 through TOC Heading 4 styles, which are included in its templates. You can change the TOC Heading level applied to a paragraph by clicking in it and then clicking on the style in the Paragraph Style pop-up menu or choosing it from the Table of Contents Styles pane that appears in the Styles drawer when you have clicked in a table of contents.

Genius

You can customize the TOC Heading styles just like any other style (see Chapter 3 for details). In particular, you may want to give a TOC Heading style a *tab leader*, a line of dots or other characters that runs from the end of the table-of-contents entry to the corresponding page number. To add a tab leader, use the Tabs pane of the Text Inspector.

Updating or deleting a table of contents

The easiest way to update a table of contents is to Ctrl+click or right-click it and choose Update Table of Contents Field. If you have the Document Inspector open, you can click the Update Now button in the TOC pane instead. Pages adds or removes any paragraphs that have changed, and updates the page numbers.

To delete a table of contents, simply click it so that Pages displays a blue border around it, and then press Delete.

Using the Proofreader

The Proofreader is Pages' feature for checking the grammar and style in your documents. Some people find it helpful, while others turn it off and leave it off. The nearby sidebar discusses the main limitations of the Proofreader to help you decide whether to use it.

To control the Proofreader, open the Edit menu, highlight Proofreading, and then click the command you want:

- **Proofreader.** Click this command to run the Proofreader, which displays the Proofreading dialog box (see figure 5.8) showing the first query it has found. Use the Correct button and the Next button to work through the queries. You can edit the suggested text before inserting it by clicking the Correct button.

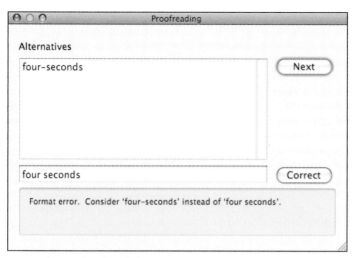

5.8 Use the Proofreading dialog box when you want to plow through the Proofreader's queries in order.

● **Proofread.** Click this command to start Proofreader proofreading. Proofreader puts a green dotted line under any query. You can Ctrl+click or right-click the queried item to see suggestions for fixing it (see figure 5.9).

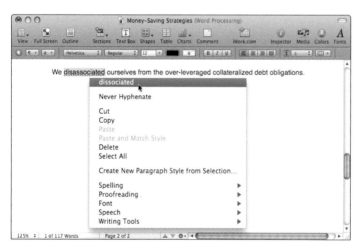

5.9 Viewing Proofreader's suggestions for fixing a problem.

● **Proofread as You Type.** Click this command to set the Proofreader to watch as you type and raise queries about grammar and style that fall afoul of its rules. Proofreader puts a green dotted line under each query.

Understanding the Limitations of the Proofreader

Computers are great for checking spelling because spelling is a simple, yes or no question: Does this group of characters match one of the words in the dictionary file? If yes, leave the word alone; otherwise, query it.

As a result, checking spelling on a computer can save you time, effort, and mistakes — so it's usually well worth doing.

By comparison, grammar and language usage are extremely flexible, which makes them almost impossible for computers to assess properly. If you doubt this, consider that English grammar is hard enough that most native speakers make mistakes every day, and variable enough to prompt bitter arguments among professional editors.

Worse yet, Proofreader isn't even a native speaker. It doesn't actually understand the meaning of your text, so its suggestions may be completely wrong. This means you must evaluate each of its suggestions against your knowledge of the language before accepting it.

Some of Proofreader's suggestions are easy to evaluate — for example, it queries repeated words and capitals that are obviously missing. You may also find its questioning of jargon and overly complex wording helpful. But for wider questions of grammar, you'll need to use your own sense of language — or a reference work.

If you're worried about the grammar in an important document, ask a colleague who has good language skills to read through it for you. Better still, ask a professional editor; even if you have to pay for his or her services, the results will be worth it.

Sharing Your Documents

When you've finished a Pages document, you'll probably want to share it with other people. You can share a document by placing it on iWork.com, by sending it to iWeb for download as either a PDF file or a Pages document, by exporting it as a Microsoft Word document, by creating a PDF file of it, or by creating a Rich Text Format (RTF) or a plain text version. Table 5.1 summarizes when to use these different options.

Table 5.1 Ways of Sharing Your Document

Document Type	Share Your Document in This Way When
Share on iWork.com	You want to make the document available for other iWork.com users to comment on or download
Share via Mail	You want to send the document via email
Send to iWeb as a PDF file	You want visitors to your iWeb Web site to be able to view the document exactly as you created it
Send to iWeb as a Pages document	You want visitors to your iWeb Web site to be able to download the document and edit it in Pages
Create a PDF file	You need to share a document in a way that retains the same layout it has in Pages
Create a Word document	You need to work with people who use Microsoft Word
Create an RTF document	You want to create a document that can be opened and edited in just about any word processor or text editor, but retains as much formatting as possible
Create a plain text document	You want to create a document that contains only the text of your document, without any formatting — for example, for sharing via text-only email

Before you share a document, open it if it's not already open. If it is open, save any unsaved changes by pressing ⌘+S or choosing File ⇨ Save.

Sharing a document on iWork.com

When you're collaborating with other users of Apple's iWork.com online service, iWork.com is a great way to share Pages documents. See Chapter 1 for a walk-through of the mechanics of the process. You can share a Pages document in any or all of four formats: Pages '09 format, Pages '08 format, Word document format, or PDF.

Sending a document via email

You can attach a Pages document to an email message just as you can any other document, but you can also quickly share it by choosing Share ⇨ Send via Mail and then selecting the format from the Send via Mail submenu: Pages, Word, or PDF.

Sending a document to iWeb for download

Pages makes it easy to send a document to a blog or podcast on your iWeb site so that visitors to the site can download it. This is a great way to distribute brochures, reports, or other documents you expect people to read offline or on their computers but not necessarily in a Web browser.

To send a document to iWeb for download, follow these steps:

1. **Choose File ⇨ Send to iWeb, and then choose the format from the submenu:**

   - **PDF.** Choose this format unless you want people to be able to edit the document in Pages.

   - **Pages.** Choose this format when the people who download the document will need to edit it using Pages.

2. **Pages opens iWeb (or activates it if it's already open).** iWeb displays the Which blog do you want to send the files to? dialog box (see figure 5.10).

3. **In the Blogs pop-up menu, choose the blog, and then click OK.** iWeb creates a page for the document. You can then fill in the placeholders as usual.

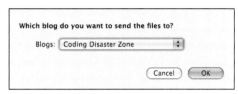

5.10 Choose which iWeb blog or podcast you want to send the Pages document or PDF to.

Creating Word documents

When you need to share a Pages document with someone who uses Microsoft Word, you can save it as a Word document with just a few clicks. But to get the best results, you may need to adjust the document in Pages before you export it.

Understanding what exports well and what doesn't

When you export a Pages document to Word, Pages does its best to produce a Word document with all the same contents and with every element in the same place. But because of the differences between Pages documents and Word documents, the Word document often isn't exactly the same.

First, the good news. These are the things that Pages gets right (or mostly right):

- **Text.** The text of your Pages document appears in the Word document in full, as it should.

- **Styles.** Pages creates paragraph styles, character styles, and list styles that Word can read. The styles are in the document itself rather than in the template to which the document is attached.

Note Pages attaches Word's Normal template to the exported document. Normal is the template Word uses for new "blank" documents (though they can include text or other items that you insert in Normal). You can attach a different template to the document by using the Templates and Add-Ins dialog box. If you select the Automatically update document styles check box in this dialog box, Word applies the styles in the template you chose to the document's existing text; if you do not select this check box, Word simply makes the template's styles available for use.

- **Tables.** Pages converts most tables successfully to Word format. However, you may need to adjust complex tables in the Word document to make them look exactly the way they did in Pages.

- **Images and shapes.** Pages includes images and shapes in the Word document. You may need to resize, reposition, or rewrap text around images or shapes.

- **Charts.** Pages can export most charts successfully as charts, so you can edit them using Word's chart tools.

Note At this writing, Word has no conversion filter for importing Pages documents. So if you need to move a document from Pages to Word, you must use Pages' export feature — and then make any adjustments needed.

When you export Pages documents to Word, you may need to adjust the following four items:

- **Page breaks.** The Word document may end up with different page breaks than those in Pages, so the document may contain more pages overall — or fewer. In manuscripts, reports, and papers, this isn't usually too much of a problem; but in a document that requires a precise layout, you may need to redo some of it. Whatever the document type, it's worth checking that the Word document doesn't contain pages that are broken in unsuitable places.

- **Column layouts.** You may find that column widths change in the exported Word document. Usually you can change them back quickly and easily.

- **Wrapping.** Text wrapped around text boxes, shapes, images, and so on in the Pages document may come out differently in Word. Similarly, paragraphs that weren't wrapped in Pages may be wrapped in Word because an object has changed position by a short distance.

- **Transparency.** If you've used the Alpha feature to make part of an image transparent in the Pages document, you'll usually need to edit the image in Word to achieve a similar effect.

If you have Word on your Mac (or on a PC that's available), usually the best way to proceed is to export the Pages document without making changes to it. Open the resulting document in Word, and see what has exported satisfactorily and what needs adjusting. You can then decide whether to fix problems in Pages and export the document again, or fix them in Word.

Genius

If you have a PC available but not Microsoft Word, download the free Word Viewer from the Microsoft Web site (www.microsoft.com/downloads). As its name suggests, Word Viewer lets you open and view Word documents, but not edit them.

Performing the export

After all those preliminaries, the export process could hardly be simpler:

1. **Choose Share ⇨ Save As to open the Export dialog box.**

2. **Click the Word button to display the Word pane.** It does not contain options you can set.

3. **Click Next to display the Save As dialog box.**

4. **Choose the name and folder for the exported document, and then click Export.**
   Pages displays a readout of its progress.

Note

You can also create a Word version of a document by choosing File ⇨ Save As, selecting the Save copy as check box, choosing Word Document in the pop-up menu, and then clicking Save.

Creating PDF files from documents

When you want to share a Pages document in its full glory — with its full text, formatting, and layout firmly in place — but make it hard for the recipient to edit the document, create a PDF file. The PDF file includes images, objects, and features such as hyperlinks (to Web sites, table of contents entries, bookmarks, and footnotes and endnotes), so it's a good way to distribute the document.

Most current operating systems include an application for reading PDF files — for example, Mac OS X includes the Preview application. If the operating system doesn't include a PDF reader, the computer's manufacturer may have installed one. If this is not the case, the recipient can download the free Acrobat Reader application from Adobe's Web site (www.adobe.com).

Genius

You can annotate a PDF document by using applications such as Mac OS X's Preview or Adobe's Acrobat Reader, but to actually edit the document, you usually need either the expensive Adobe Acrobat itself or an equivalent PDF editor.

165

Here's how to create a PDF file from a Pages document:

1. **Choose Share ⇨ Export to open the Export dialog box.**

2. **Click the PDF button to display the PDF pane (see figure 5.11) if one of the other panes is displayed.**

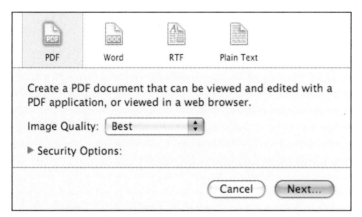

| PDF | Word | RTF | Plain Text |

Create a PDF document that can be viewed and edited with a PDF application, or viewed in a web browser.

Image Quality: [Best ▲▼]

▶ Security Options:

(Cancel)   (Next...)

5.11 The PDF pane of the Export dialog box lets you choose which quality to use for images in the PDF file.

3. **In the Image Quality pop-up menu, choose the image quality you want: Good, Better, or Best.**

   ● Best is normally what you'll want, because it produces a full-quality PDF. Pages keeps the images at their full resolution.

   ● If you find that Best produces files that are too large for the method you're using to distribute them, experiment with the Better setting or the Good setting to produce a smaller file. Better reduces the image quality to 150 dots per inch (dpi); Good uses 72 dpi.

4. **If you want to secure the PDF file, click the Security Options disclosure triangle to reveal the security options (see figure 5.12).** You can then require a password to open the document or a (different) password to print it or copy information from it.

Caution   Entering a password required to open the PDF document is usually effective. However, entering a password for copying and printing the document does not always work with every PDF reader: even if the user enters the right password, he or she may be unable to copy or print the document.

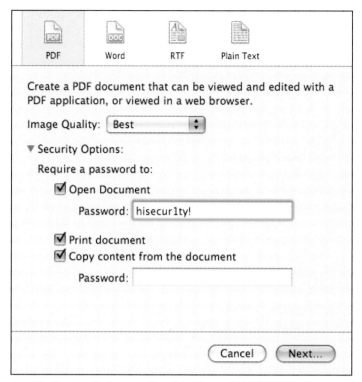

5.12 The Security Options section of the Export dialog box lets you require a password for opening the PDF, printing it, or copying content from it.

5. **Click Next.** Pages displays a Save dialog box that lets you choose the name to give the PDF and the folder in which to save it.

6. **Click Export.** Pages displays a progress readout as it exports the PDF.

After Pages finishes the export, open a Finder window to the folder in which you saved the PDF. Click the file, check its file size, and then double-click the file to open it in Preview (or your default PDF viewer). Make sure that the document appears the way you want it, and that any security options you chose are working effectively, before you distribute the file.

Creating Rich Text and Plain Text documents

If you need to create a version of a Pages document that contains only the text, export the document to a plain text document. *Plain text* means that the document contains nothing but text: It contains no formatting and no images or other objects (such as charts). To create a plain text file, Pages strips out all the objects and formatting information, leaving only the text.

Genius

You can open a plain text file in any text editor (from the simple Notepad application on Windows upward) or word-processing application. You can also open it in most email applications and other applications that can handle text input, from spreadsheet applications to database applications.

Plain text is stripped down to the bone. If you want a file that contains most of the Pages document's formatting and images, you can create an RTF document instead. An RTF document contains text with formatting and also includes images.

To create an RTF or plain text document, follow these steps:

1. **Choose Share ⇨ Export to open the Export dialog.**

2. **Click the RTF button or the Plain Text button, as appropriate.** Pages displays the RTF pane or the Plain Text pane. Neither pane contains any options.

3. **Click the Next button.** Pages displays a Save dialog box.

4. **Type the filename, and choose the folder in which to store the file.** Pages automatically assigns the file extension .rtf to an RTF file and the file extension .txt to a plain text file.

5. **Click Export.**

For either file type, open a Finder window to the folder you used, and then double-click the file to open it in TextEdit (or your default application for text files if you're using a different application). For a text file, make sure that the text is all present and in the order you expect (the contents of text boxes may come out in surprising places). For an RTF file, check the formatting and objects (such as images), as well.

How Can I Work More Efficiently in Numbers?

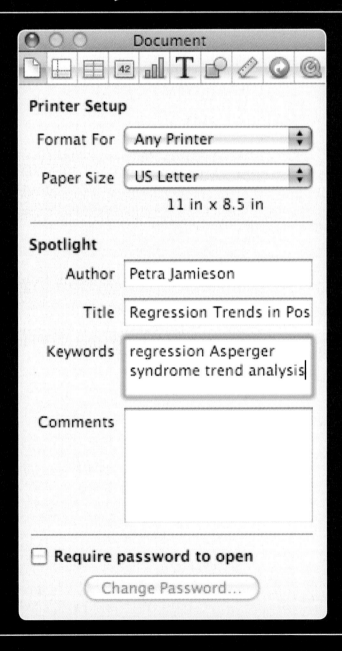

1 | 2 | 3 | 4 | 5 | 6 | 7 | 8 | 9 | 10 | 11 | 12

Numbers enables you to create powerful, functional, and attractive spread-sheets. To work as quickly and efficiently as possible in Numbers, you need to understand Numbers' components and what they do, set Numbers-specific preferences to suit the way you work, and choose which of Numbers' tools to display while you're working. You also need to know how to organize your information smartly and easily using sheets and tables, how to use Numbers' shortcuts for entering text quickly into a table, and how to import data from both Microsoft Excel workbooks and from your Mac's Address Book.

Knowing What You Are Working With

First, let's meet the various elements in the Numbers window and make sure you know what they do. After you open Numbers and pick the template you want in the Template Chooser window (assuming it appears), the Numbers window appears, as shown in figure 6.1.

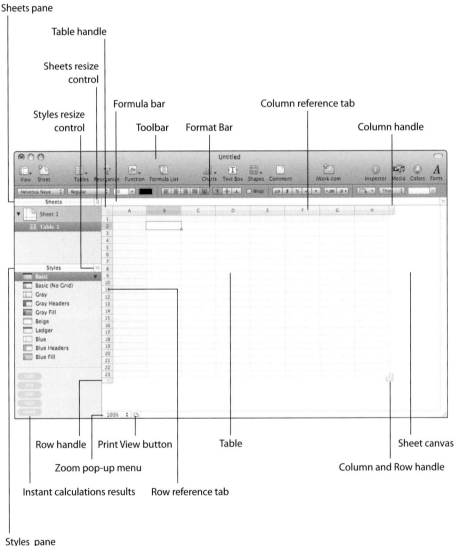

6.1 The Numbers window with a new spreadsheet open in it.

Here's what the various elements in the Numbers window are and what they do:

- **Toolbar.** Provides buttons and pop-up menus for common actions and for displaying other tools (such as the Inspector window and Media Browser). You can customize the toolbar as discussed in Chapter 1.

- **Format bar.** Gives you access to the most widely used formatting commands, such as font formatting, alignment, number formatting, and border formatting.

- **Formula bar.** This bar appears below the Format bar and lets you quickly create and edit formulas. For heavier-duty formula work, you can instead use the Formula Editor, which is discussed in Chapter 7.

- **Sheets pane.** This pane lets you choose the sheet, table, or chart you want to view, add and delete items, or rename or rearrange items. You can drag the Sheets resize control right or left to make the Sheets pane wider or narrower.

- **Styles pane.** This pane makes it easy to apply predefined formatting to cells or tables, just as you can apply styles to paragraphs or characters in Pages. You can drag the Styles resize control up or down to increase or decrease the depth of the Styles pane; the depth of the Sheets pane decreases or increases by the same amount.

- **Instant calculation results.** This area automatically calculates several widely used calculations (sum, average, minimum, maximum, and count) when you select two or more cells in a table. You can either use this information as a handy point of reference or drag the formula for one of the calculations to a cell in which you want to use it.

- **Zoom pop-up menu.** Use this menu to zoom the view in, so that the contents of the sheet appear larger, or out, so that you can see more of the sheet at once.

- **Print View button.** Click this button to switch between Normal view and Print view.

- **Table handle.** Click this button to select the whole table so that you can move it.

- **Column handle.** Drag this handle to the right to reveal more columns or to the left to hide some of the displayed columns.

- **Row handle.** Drag this handle down to reveal more rows or up to hide some of the displayed rows.

- **Column and Row handle.** Drag this handle down and to the right to reveal more columns and rows at the same time. Drag the handle up and to the left to hide some of the displayed columns and rows.

Setting Numbers-Specific Preferences

To make Numbers behave the way you want, you can set various preferences. Apart from the iWork-wide preferences discussed in Chapter 1, Numbers also includes several General preferences that Pages and Keynote do not have. These are discussed here.

You can open the Preferences window in either of the standard Mac OS X ways:

- **Mouse.** Choose Numbers ⇨ Preferences.
- **Keyboard.** Press ⌘+, (comma).

Choosing Formulas preferences

The Formulas area contains two preferences that can make a huge difference to your work in Numbers:

- **Use header cell names as references.** Select this check box to refer to a cell by the names of its column header and row header (in that order). For example, in a table that has a column named 2010 and a row named Indianapolis, you can refer to the cell at their intersection by using "2010 Indianapolis" instead of its standard name (for example, A15). It can be faster and more convenient to use header cell names, so you'll probably want to experiment with it.

- **Show warnings when formulas reference empty cells.** Select this check box if you want Numbers to display a blue warning triangle in the cell of any formula that includes a reference to an empty cell. When you're finishing up a spreadsheet, this warning can be helpful, but if you're developing your own spreadsheets, you may find that so many warnings become distracting. If so, turn the warnings off until the spreadsheet is mostly complete, then turn them back on when you need to track down errors.

Choosing Currencies and Editing preferences

In the Currencies area, select the Show complete list of currencies in Cells Inspector check box if you want to be able to choose from a full list of currencies rather than just the most widely used ones.

If you're working with widely used currencies such as the US dollar, pound sterling, yen, won, or rupees, stick with the short list: It includes all the major currencies and makes it easier to find the one you need.

Caution

If a column contains many entries that start with the same letters or word, the auto-completion list may slow down your data entry. If this happens, deselect the Show auto-completion list in table columns check box in Numbers preferences.

In the Editing area, you can select the Automatically move objects when tables resize check box to enable Numbers to move objects (such as shapes and images) as needed out of the way of tables when you resize a table. For example, if you add rows or columns to the table, Numbers may need to move objects. Deselect this check box if you prefer to move objects yourself.

Customizing the Numbers Window for Faster Work

To work fast and effectively in Numbers, spend a few minutes setting up the application's window to suit the way you work. That means zooming the window to a comfortable size, picking the right view for the task you're doing, and choosing which screen elements to display.

Zooming to a comfortable size

The first essential is to zoom the Numbers window in or out to a size at which you can comfortably work. To do so, open the Zoom pop-up menu at the bottom-left of the sheet canvas and click the size you want.

Genius

You can also zoom quickly using the keyboard. Press ⌘+> to zoom in by one of the zoom increments on the pop-up menu or ⌘+< to zoom out by one zoom increment. You can also choose View ➪ Zoom ➪ Zoom In or View ➪ Zoom ➪ Zoom Out, but usually the keyboard shortcuts are quicker. To zoom the sheet to its actual size, choose View ➪ Zoom ➪ Actual Size (there's no keyboard shortcut for this command).

The best size depends on your monitor, your display resolution, and your eyes. Take a minute to experiment with different zoom levels to find out which is best for you.

Switching among views

Most of the time, you'll use Numbers' Normal view, which displays the current sheet without headers, footers, or page breaks, letting you concentrate on your tables, charts, and other objects.

When you want to see how your spreadsheet will look when divided into pages, switch to Print view by clicking the Print View button at the bottom of the Numbers window or choosing Show Print View from either the View pop-up menu on the toolbar or the View menu on the menu bar. Chapter 9 explains how to make the most of Print view.

To switch back from Print view to Normal view, click the Print View button again or choose Hide Print View from either the View pop-up menu or the View menu.

Genius

You can also press ⌘+Option+P to switch to or from Print view.

Choosing which screen elements to display

Numbers lets you display or hide several parts of its window to suit the spreadsheet you're working on and the changes you're making to it. You can choose to display or hide the toolbar, Format bar, rulers, Formula list, and comments, as shown in figure 6.2.

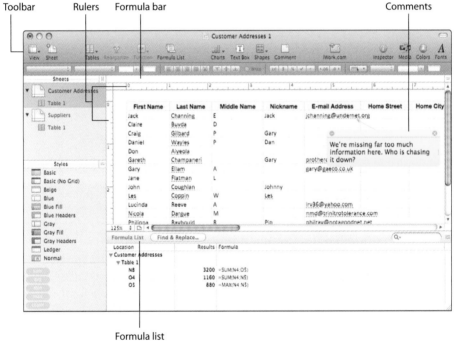

6.2 You can choose to display or hide parts of the Numbers window.

Toolbar

The toolbar across the top of the Numbers window provides buttons for frequent actions, such as choosing which screen elements to display, adding sheets, and opening the Inspector window. You can customize the toolbar with your favorite buttons as discussed in Chapter 1.

You can also hide the toolbar if you need more screen space. To display or hide the toolbar:

- **Mouse.** Click the jellybean-shaped button at the right end of the title bar.
- **Keyboard.** Press ⌘+Option+T.
- **Menu bar.** Choose View ⇨ Hide Toolbar or View ⇨ Show Toolbar.

Format bar

The Format bar appears under the toolbar (when the toolbar is displayed) and provides quick access to widely used formatting commands.

You can display or hide the Format bar in these ways:

- **Mouse.** Click the View pop-up menu on the toolbar and choose Show Format Bar or Hide Format Bar.
- **Keyboard.** Press ⌘+Shift+R.
- **Menu bar.** Choose View ⇨ Hide Toolbar or View ⇨ Show Toolbar.

Rulers

You don't need to see rulers most of the time you're working on spreadsheets, so Numbers doesn't display the horizontal and vertical rulers at first. When you need the rulers, you can display and hide them in these ways:

- **Mouse.** Click the View pop-up menu on the toolbar and choose Show Rulers or Hide Rulers.
- **Keyboard.** Press ⌘+R.
- **Menu bar.** Choose View ⇨ Show Rulers or View ⇨ Hide Rulers.

Formula list

The Formula list is a pane that you can open at the bottom of the Numbers window to review a list of the formulas in the spreadsheet. Each sheet and table appears as a list that you can collapse or expand by clicking its disclosure triangle, enabling you to focus on the formulas you're interested in. You can also search for formulas and use Find & Replace to make sweeping changes to them (click the Find & Replace button in the Formula list).

You can display or hide the Formula list in these ways:

- **Mouse.** Click the View pop-up menu on the toolbar and choose Show Formula List or Hide Formula List.
- **Keyboard.** Press ⌘+Option+F.
- **Menu bar.** Choose View ⇨ Show Formula List or View ⇨ Hide Formula List.

Comments

Numbers lets you attach a comment to a cell, which is useful when you need to explain what's happening in a spreadsheet or when you're reviewing someone else's spreadsheet.

Numbers normally displays comments in a spreadsheet, but you can hide them by choosing Hide Comments from either the View pop-up menu on the toolbar or the View menu on the menu bar. Choose Show Comments from either of these menus when you want to view the comments again.

Organizing Your Information with Sheets and Tables

Each Numbers document consists of one or more worksheets, or simply sheets, on which you can place tables and other objects as needed. The Sheets pane lists the sheets and tables in the document and lets you easily manage them. The current sheet appears on the sheet canvas, the main work area in the Numbers window.

Adding and deleting sheets

You can quickly add a sheet in any of these ways:

- Click the Sheet button on the toolbar.
- Choose Insert ⇨ Sheet.
- Ctrl+click or right-click in the Sheets pane and choose New Sheet.

To delete a sheet, click it in the Sheets pane, and then either choose Edit ⇨ Delete Sheet or press Delete. Alternatively, Ctrl+click or right-click the sheet in the Sheets pane and choose Delete Sheet.

To avoid accidentally deleting vital data, Numbers double-checks that you want to delete the sheet (see figure 6.3). Click the Delete button.

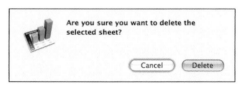

6.3 When you delete a sheet, Numbers prompts you to confirm the decision.

Moving and copying sheets

You can rearrange the sheets in the Sheets pane into the order you want. Simply drag a sheet to where you want it to appear.

To move a sheet from one spreadsheet document to another, cut it from the source spreadsheet and paste it into the destination spreadsheet. For example, Ctrl+click or right-click the sheet in the source Sheets pane and choose Cut, then Ctrl+click or right-click in the destination Sheets pane and choose Paste. You can also press ⌘+X to cut from the keyboard and ⌘+V to paste, as usual in Mac OS X.

Often, you can save time by copying a sheet and then making any changes the copy needs.

To copy a sheet within a spreadsheet document, simply Option+drag the sheet up or down in the Sheets pane. You can also Ctrl+click or right-click the sheet and choose Copy (or select the sheet and press ⌘+C), then Ctrl+click or right-click again and choose Paste.

Numbers adds "-1" to the end of the copy's name to distinguish it from the original. You can then rename the sheet to whichever name you prefer, as described in the next section.

To copy a sheet from one spreadsheet document to another, copy it from one Sheets pane and paste it into the other Sheets pane. The copied sheet keeps its own name unless another sheet already present uses that name (in which case Numbers adds "-1" to the name).

Renaming a sheet

Numbers names the first sheet in a spreadsheet Sheet 1 and gives the next available number to each sheet you add: Sheet 2, Sheet 3, and so on. As soon as the spreadsheet is taking shape, give each sheet a descriptive name to make it easier to identify.

To rename a sheet, double-click the name in the Sheets pane, type the name in the edit box that appears, and then press Return. You can also display the edit box by Ctrl+clicking or right-clicking the name and choosing Rename.

Note You can also change a sheet's name in the Name box of the Sheet Inspector.

Normally, it's best to use names short enough to fit in the Sheets pane at your preferred width for it. You can drag the Sheets resize control to change the width of the Sheets pane as needed, but there's little point in sacrificing a large chunk of the Numbers window just to accommodate long names.

Genius Each sheet works as a single page, but you can divide it up as necessary. For example, you can set a sheet to print on multiple pages of paper. Chapter 9 explains how to do this.

Selecting parts of a table

To work with parts of a table, you need to select them. Here's how to do so:

- **Select a cell.** Click the cell.

- **Select a range of adjacent cells.** Click the first cell and drag through the rest. Alternatively, click the first cell, and then Shift+click the last cell.

- **Select a column or row.** Click the column reference tab or row reference tab.

- **Select the whole table.** Click the Table handle in the upper-left corner. You can also press ⌘+Return when any part of the table is selected.

Resizing and moving a table

If your table is the wrong size, click the Table handle to display selection handles around the table. You can then click and drag a side handle to change the table's width, a top or bottom handle to change its height, or a corner handle to change both.

You can also resize a table by adding or deleting rows or columns, as explained a little later in this chapter.

If the table is in the wrong place on the sheet canvas, click the Table handle, move the mouse pointer over a table border so that it displays four directional arrows, and then drag the table to where you want it.

Naming a table

Each table in a sheet must have a unique name, so Numbers automatically names the tables Table 1, Table 2, and so on. When you're ready to rename a table, double-click the name in the Sheets pane, type the new name in the edit box that appears, and then press Return. You can also Ctrl+click or right-click the name and choose Rename to open the edit box.

You can also change a table's name in the Name box in the Table Inspector.

Adding rows or columns to a table

Dragging a sizing handle down, right, or both is a convenient way of adding rows or columns to the end of a table, but you'll often need to add rows or columns in the middle of the table. You can do so in three ways:

- **Menus.** Click anywhere in the table and then open the Table menu or Ctrl+click or right-click in a table to produce the shortcut menu. You can then choose the Add Row Above, Add Row Below, Add Column Before, or Add Column After command.

● **Keyboard shortcuts.** Press Option+Up Arrow to add a row before the current row, or Option+Down Arrow to add a row after it. Press Option+Left Arrow to add a column before the current column, or Option+Right Arrow to add a column after it.

● **Reference tab pop-up menu.** Move the mouse pointer over a column reference tab or row reference tab, and then click the pop-up button that appears. From the column reference tab pop-up menu (see figure 6.4), you can choose Add Column Before or Add Column After. From the row reference tab pop-up menu, you can choose Add Row Above or Add Row Below.

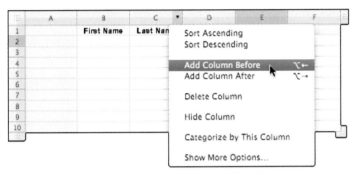

6.4 The reference tab pop-up menus let you quickly add columns or rows to a table.

Rearranging rows and columns

When you get your rows or columns in the wrong places, you can quickly sort things out.

To move a row or a column a short distance, click and drag its reference tab to where you want the row or column to appear. Numbers displays a double blue line to show where the item you're dragging will land (see figure 6.5).

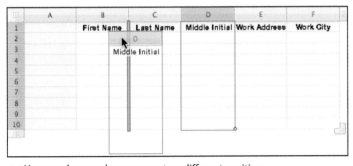

6.5 You can drag a column or row to a different position.

To move a row or column farther, use the Mark for Move command like this:

1. **Click the header for the row or column to select the row or column.**

2. **Choose Edit ⇨ Mark for Move.** Numbers displays a brown outline around the row or column.

3. **Move to the destination, and then click the row header or column header.**

4. **Choose Edit ⇨ Move.** Numbers moves the row or column to the destination.

Deleting rows and columns

You can delete a row or column either of these ways:

- Open the reference tab pop-up menu for the column or row, and then choose Delete Column or Delete Row.

- Click in a cell in the column or row, and then choose Edit ⇨ Delete Column or Edit ⇨ Delete Row. The Delete Column command and Delete Row command also appear on the Table menu.

Hiding rows and columns

Often, it's useful to hide rows or columns from view — for example, so that they're not visible to someone with whom you're sharing the spreadsheet, or simply so that you can see all the relevant columns or rows at once.

To hide a row or column, click its reference tab, and then choose Hide Row or Hide Column from the menu. Numbers gives no visual indication that the row or column is hidden except that the reference tabs no longer show its number or letter. For example, if you hide column C, the reference tabs show A, B, D, E, and so on.

To hide multiple rows or columns in a single move, select one or more cells in each row or column. Click a reference tab for one of those rows or columns and choose Hide Selected Rows or Hide Selected Columns.

You can reveal the hidden items again using any of these methods:

- **Reveal one or more hidden rows or columns.** Highlight the reference tab for the row or column before or after the hidden rows or columns, open the reference tab's menu, and then choose the Unhide command — for example, Unhide Column C or Unhide Rows 4–7.

● **Reveal separate hidden rows or columns.** Select a range of cells in the rows or columns that include the hidden rows or columns. Open one of the reference tab menus and choose Unhide Selected Rows or Unhide Selected Columns.

● **Reveal all hidden rows or columns.** Choose Table ⇨ Unhide All Rows or Table ⇨ Unhide All Columns.

Adding table header rows, header columns, or footer rows

To make a table clear to read, you'll usually need to give it one or more header rows, header columns, or footer rows. Numbers lets you add up to five of each of these. Normally, one or two is enough, but it's useful to be able to add more when you need them.

To quickly add or remove a header row, header column, or footer row, click anywhere in the table, open the Header columns pop-up menu, the Header rows pop-up menu, or the Footer rows pop-up menu at the right end of the Format bar (see figure 6.6), and choose the number of rows or columns: 0, 1, 2, 3, 4, or 5.

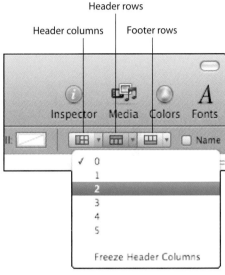

Header rows

Header columns

Footer rows

6.6 You can quickly add a table header row, header column, or footer row from the Format bar.

Note

If the Numbers window isn't wide enough to display the Header columns pop-up menu, the Header rows pop-up menu, and the Footer rows pop-up menu, open the Table menu and use the Header Columns submenu, the Header Rows submenu, or the Footer Rows submenu instead.

Adding more header columns, header rows, or footer rows from the reference tab

You can also add header columns, header rows, or footer rows from the reference tab pop-up menu for an existing header or footer. For example, highlight the reference tab for an existing header column, open the pop-up menu, and choose Add Header Column Before or Add Header Column After.

Freezing a table's header columns or header rows

When you've added headers to a table, you'll often want to "freeze" the columns or rows so that they don't disappear off the screen when you scroll. Having the headers remain in place lets you identify the contents of cells more easily.

To freeze a header column or header row, click in the table, open the Header columns pop-up menu or Header rows pop-up menu on the Format bar, and choose Freeze Header Columns or Freeze Header Rows. (If these pop-up menus don't appear on the Format bar, open the Table menu and choose the commands there instead.) When there's a check mark next to the Freeze option, the headers are frozen and won't disappear when you scroll.

You can't freeze footer rows.

Note

Bringing in a table from Pages

If you've created a table in Pages, you can bring it into Numbers by copying it from the Pages document and then pasting it onto a sheet in Numbers. Numbers picks up all the formatting that you've applied, together with any formulas you've entered in the table.

Entering Text in a Spreadsheet

You can type or paste data into a table, but Numbers also provides handy features for entering series of items and identical items quickly. Numbers also makes it easy to enter dates and hyperlinks in your tables, and to apply cell formatting to make cell contents appear the way you want them to.

Entering text quickly with autofilling

Many tables need predictable sequences of data in adjacent cells. For example, you may need to enter a sequence of years in a column or the months of the year in a row, or you may need to enter increasing values (such as 5, 10, 15, and so on) in a series of cells.

To help you enter such data quickly, Numbers provides a feature called autofilling. You select one, two, or more cells that contain the information required to start the series, and then click and drag the autofilling handle in the lower-right corner of the last cell to tell Numbers which cells you want to fill with the data.

The autofilling handle is the little white diamond that appears in the lower-right corner of the active cell. When you move the mouse pointer over the autofilling handle, the pointer changes to a heavy black cross to tell you it's in the right place.

The best way to learn to use autofilling is to experiment with it. Open a practice spreadsheet and try these examples:

- **Click a cell, type January, and then click and drag the autofilling handle across to the right.** Numbers fills in a month for each cell you drag, as shown in figure 6.7.

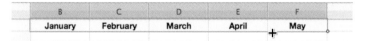

6.7 Click and drag the autofilling handle to make Numbers automatically fill a series of data into the cells you select.

- **Click a cell and type 10.** Click and drag the autofilling handle to the right. Numbers enters 10 in each cell, because there's no progression.

- **Click the cell below the 10 cell and type 20 in it.** Select the two cells, and drag the autofilling handle down from the second cell. Numbers fills in 30, 40, 50, and so on for you, because the first two cells show the number increasing by 10 from 10.

Genius

You can sometimes use autofilling on two or more different series of data at once. For example, if you type January in cell B1 and 2010 in cell B2, you can select both cells and drag the autofilling handle to the right. Numbers fills in February 2010 in cells C1 and C2, March 2010 in cells D1 and D2, and so on.

- **Click a cell and type Monday in it.** Select that cell and the six cells below it. Then choose Insert ⇨ Fill ⇨ Fill Down to make Numbers fill those six cells with the other days of the week. (You can also use the Fill Up command when you've selected cells in a column, or use the Fill Left command or the Fill Right command when you've selected cells in a row.)

Entering dates

Dates can be a real headache for humans to calculate, but Numbers makes them as painless as possible.

You can type a date in a wide variety of formats and have Numbers recognize it. Here are examples of what Numbers will recognize:

- 12/13, 12-13, Dec 13, or 13 Dec

Note

If you omit the year when you enter a date, Numbers assumes you mean the current year. Numbers adds that year to complete the date.

- 12/13/09, 12-13-09, 13 December 09, 13 Dec 09
- Dec 13, 2009 (with the comma)
- 12/13/2009, 12-13-2009, 13 December 2009

When you enter a date in a cell formatted with the Automatic cell format, Numbers automatically displays the date in a standard date format. For example, Numbers typically displays the example date shown here as Dec 13, 2009.

You can customize the way that Numbers displays dates, as discussed later in this chapter in the section on formatting cell values.

Note

If you're used to Excel, you may find Numbers' way of calculating times tricky. Unless you adjust the cell formatting, Numbers assumes you're entering times in a week-day-hour-minute-second format (for example, 1 week 2 days 3 hours 4 minutes and 5 seconds). When you need to make Numbers calculate only the time, format the cell as hour-minute-second format.

Entering hyperlinks

For reference, it's often useful to enter hyperlinks in your tables. Numbers lets you create either a hyperlink that opens a browser to a Web address (URL) or starts a new email message to an email address.

Here's how to insert a hyperlink:

1. **Double-click the cell in which you want to insert the hyperlink.** If the cell contains text that you want to use for the display name of the hyperlink, select that text.

2. **Choose Insert ⇨ Hyperlink, and then choose the type of hyperlink from the Hyperlink submenu: Webpage or Email Message.** Numbers inserts a standard hyperlink of the type you chose, opens the Inspector window if it's closed, and displays the Hyperlink Inspector (see figure 6.8).

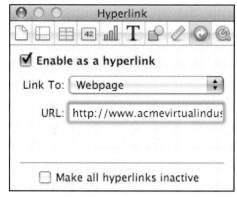

3. **Enter the address for the hyperlink.**

   ● For a Web address, enter the URL in the URL box. You can type it if you want, but for any complex address, copying it from a Web browser and pasting it in is usually a better bet.

6.8 Enter the URL or email address for the hyperlink in the Hyperlink Inspector. To see the full URL, hold the mouse pointer over the URL box for a second.

   ● For an email message, enter the address in the To box and type the default subject in the Subject box. The sender can change the subject as needed, but entering a default one reduces your chance of getting messages with blank subject lines.

4. **Make sure the Enable as a hyperlink check box is selected.** Numbers selects this check box automatically when you create a hyperlink, but you can deselect it if you find yourself clicking the hyperlink by accident.

Genius

When you're creating a table, you may want to prevent any of the hyperlinks from triggering. To do so, select the Make all hyperlinks inactive check box in the Hyperlink Inspector. Deselect this check box when you're ready to turn the hyperlinks back on.

Formatting cell values

To make your tables look right, you'll need to format the cells to suit their contents. For example, you may need to change the number of decimal places shown for numbers, change the currency symbol shown for money, or choose a different date format.

You can quickly apply widely used formatting using the controls on the Format bar, or you can open the Cells Inspector and use the options in the Cell Format area.

Understanding the Automatic format

In many of the templates that come with Numbers, most of the cells are formatted using the Automatic format. This format tries to work out which type of data you've entered in the cell and format it in a suitable way.

For example, if you type a number in a cell, Numbers keeps the commas and decimal places as you enter them. If you type a currency value, Numbers displays two decimal places (for example, $75.25) unless the number is an integer, in which case it displays no decimal places (for example, $75). And if you enter "true" or "false," Numbers assumes you're using a Boolean value, and converts the word to TRUE or FALSE. (A Boolean value is one that can only be TRUE or FALSE — not other values.)

The Automatic format works well much of the time, but when you want a cell to appear in a particular way, you can apply exactly the format needed.

Applying a format from the Format bar

The quick way to apply cell formatting is to use the Format bar (see figure 6.9):

- The four buttons on the left give you quick access to the most widely used formatting options: Number with two decimal places (for example, 4.22), Currency (for example, $44.23), Percentage (for example, 10%), and Checkbox (which displays a check box that the user can select or deselect).

- The Format pop-up menu lets you reach the other formatting options, from Automatic and Duration through to Custom.

- The two buttons to the right of the Format pop-up menu let you quickly increase and decrease the number of decimal places.

Number with two decimal places button

Decrease the number of decimal places button

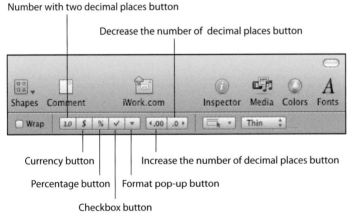

Currency button

Percentage button   Format pop-up button

Increase the number of decimal places button

Checkbox button

6.9 The Format bar provides widely used cell formatting options.

For example, to format a cell to show a number with four decimal places, click the Number with two decimal places button, and then click the Increase the number of decimal places button twice.

Applying a format with the Cells Inspector

For greater control over the formatting, open the Cells Inspector (see figure 6.10): Click the Inspector button on the toolbar, and then click the Cells Inspector button.

Open the Cell Format pop-up menu, choose the format you want, and then choose options for the format. The options available depend on the format you choose.

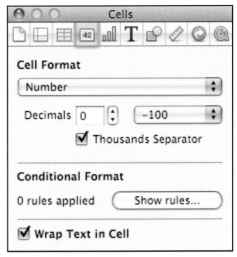

6.10 Use the Cells Inspector to choose exactly how to format a cell.

Note See Chapter 4 for a detailed breakdown of the Number, Currency, Percentage, Date & Time, Duration, Fraction, Numeral System, Scientific, Text, and Custom formats. See Chapter 8 for instructions on using the four controls that Numbers lets you put in cells: Checkbox, Stepper, Slider, and Pop-up Menu.

Using Auto-Completion lists

Numbers' Auto-Completion feature can save you time when a table column contains multiple instances of the same entry. For example, say you're creating a table that lists sales by your company's various offices. If the table has a column showing the city names, having Auto-Completion pop up a short list of the offices starting with the letter you typed can speed up data entry.

Other times, though, you may find Auto-Completion unhelpful. For example, when you're entering phone numbers for different customers, having Auto-Completion offer you every other phone number in the same area code is no use. In this case, turn Auto-Completion off by opening the General preferences (choose Numbers ➪ Preferences) and deselecting the Show auto-completion list in table columns check box.

Making Your Documents Easier to Find with Spotlight

To make your Numbers spreadsheets easier to find with Mac OS X's Spotlight search feature, you can add metadata such as an author name, a title, keywords, and comments to them. To do so, click the Inspector button in the toolbar, click the Document Inspector button, and then work in the Spotlight area. For each spreadsheet, you can add an author name, a title, keywords (which you can use to search for matching documents), and comments.

Importing Data from Microsoft Excel

Numbers is one of the newer spreadsheet applications to enter the spreadsheet arena, and as such, it has to get along with Microsoft Excel, the heavyweight champion. Numbers can import Excel workbooks and export its own documents in Excel format.

Note

Numbers can open workbooks in the Excel 2007 (Windows) and Excel 2008 (Mac) documents formats, which use the .xlsx file extension, as well as workbooks in the formats used by Excel 2003 (Windows) and Excel 2004 (Mac) and earlier versions, which use the .xls file extension.

To work with an Excel workbook in Numbers, simply open it as you would any other spreadsheet: Press ⌘+O or choose File ⇨ Open to display the Open dialog, select the Excel workbook, and then click the Open button. From the Finder, you can Ctrl+click or right-click the Excel workbook and choose Open With ⇨ Numbers from the shortcut menu.

Numbers tries to convert all the Excel workbook's contents to their equivalents in Numbers. This process normally takes a few seconds for a small workbook and can take several minutes for a large and complex one (depending on how brawny your Mac is).

Genius

You can't open an Excel workbook in Numbers, edit it there, and then save your changes back to the original Excel file. (Some other spreadsheet applications let you do this.) Instead, you must save the file as a Numbers spreadsheet; if you want an Excel version of that spreadsheet, you can export a version of the spreadsheet to Excel format (see Chapter 9 for instructions).

If you open only modest-size Excel workbooks that contain straightforward data and formulas, Numbers can usually open them without any problem. But if you have larger or more complex Excel workbooks, Numbers may struggle to open them.

Here are things that tend to go wrong when importing Excel workbooks, in approximate order of severity:

- **Numbers can import only 255 columns and 65,533 rows.** If you open an Excel workbook that has more rows or columns, Numbers removes the additional rows and gives you an Import Warning in the Document Warnings dialog box.

- **Numbers can't handle huge Excel workbooks.** If you see the Import Error dialog box shown in figure 6.11, claiming "The document can't be imported because it's too large," you've run headfirst into this problem. This problem isn't the number of rows or columns, but simply the amount of data in the workbook. There's no specific file size that's too big, as it depends on how much RAM your Mac has, how many other applications it's running when you take on the heavyweight workbook, and how complex the data in the workbook is. But you'll find that workbooks that Excel for the Mac can handle without breaking a virtual sweat can cause Numbers to founder — on the same Mac.

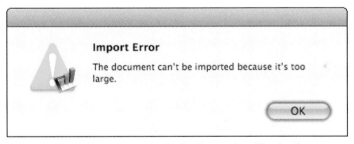

Import Error

The document can't be imported because it's too large.

OK

6.11 Numbers has trouble importing chunky Excel workbooks. If you run into this error, there's nothing to do but click OK, even if it's not okay. Numbers then gives up on opening the workbook.

- **Excel worksheet headers and footers get mangled.** Excel encourages you to set up worksheet headers and footers that have a left section, a center section, and a right section. Numbers imports these sections as separate paragraphs in the header rather than as items aligned left, center, and right on the same line. You can fix this problem in a few seconds, but it can be a killer if you need to open many Excel spreadsheets that use such headers and footers.

- **You may need to change some formulas.** For example, if you've used the SUMIF function in Excel, you may need to use the SUMIFS function in Numbers to get the same result. Also, see the next point.

191

- **Numbers doesn't support all Excel's functions — not by a long shot.** Excel has around 500 functions, while Numbers '09 has 261 functions. The 239-odd functions that Numbers doesn't have are mostly specialized ones, but if you work with statistics, it's a good idea to check that Numbers provides the functions you need. For a full list of Numbers' functions, download the *iWork Formulas and Functions User Guide* from the Support area of the Apple Web site.

- **Numbers doesn't offer Excel's drop-down lists.** In Excel, you can create a list of pre-defined values — for example, a list of your company's offices or of project codes — and then use a drop-down list to enter them in cells. Numbers simply doesn't offer this fea-ture. The best workaround is usually to store the predefined values in an ancillary table in your Numbers document, and then use a VLOOKUP function to enter the data in the main table.

- **Some formatting disappears.** For example, Excel lets you rotate text within a cell, so if you need text to run at a 45-degree angle, all you need to do is apply the formatting. Numbers strips out any formatting it can't handle. To rotate text in Numbers, insert a text box, rotate it, and then enter the text.

- **Locking and password protection.** Excel lets you lock cells against editing and protect a worksheet with a password. Numbers doesn't support these features, so it strips out the locking and protection when you open an Excel workbook. Numbers warns you about this change in the Document Warnings dialog box.

Importing Data from Address Book

Numbers lets you import data directly from your Mac OS X Address Book. This is great when you need to put together tables showing customer contact information, guest addresses, or other data you can pull from your contacts.

Knowing which fields you can use

You can use all the fields that Address Book offers except for the custom fields that you add your-self. The easiest way to use the fields is simply to give your Numbers table the same names as the fields — for example, create column headers called Prefix, First Name, Last Name, and so on. However, Numbers also lets you use synonyms for some of the most widely used fields in Address Book. Table 6.1 explains the synonyms you can use.

Table 6.1 Numbers' Synonyms for Address Book Field Names

Address Book Field Name	Synonyms
Prefix	Name Title, Name Prefix
Last Name	Last, Surname
First Name	First, Given Name, Forename
Department	Job Department
Mobile	Mobile Phone, Mobile Telephone, Cell Phone, Cell Telephone, Cellular, Cellular Phone, Cellular Telephone
Pager	Beeper
Email	Email Address
AIM	IM, IM Handle, IM Name, IM Address, Chat, Chat Handle, Chat Name, Chat Address
Work AIM	Work IM, Work IM Handle, Work IM Name, Work IM Address, Work Chat Handle, Work Chat Name, Work Chat Address
Home AIM	Home IM, Home IM Handle, Home IM Name, Home IM Address, Home Chat Handle, Home Chat Name, Home Chat Address
Other AIM	Other IM, Other IM Handle, Other IM Name, Other IM Address, Other Chat Handle, Other Chat Name, Other Chat Address
Street Address	Street
City	Town
Zip	Zip Code, Postal Code
Work Street Address	Work Street, Work Address
Work City	Work Town
Work Zip	Work Zip Code, Work Postal Code
Home Street Address	Home Street, Home Address
Home City	Home Town
Home Zip	Home Zip Code, Home Postal Code
Other Street Address	Other Street, Other Address
Other City	Other Town
Other Zip	Other Zip Code, Other Postal Code
Note	Notes

Importing the data into a table

Here's how to import data from Address Book into a table:

1. **Open Address Book so that you can see the field names.**

2. **In Numbers, create a table with a header row that contains the field names you want to use.** Figure 6.12 shows an example of a simple table with the headers First Name, Last Name, and Work Phone.

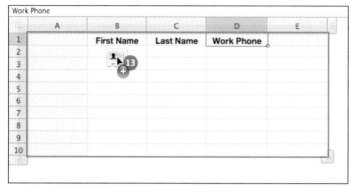

6.12 You can drag contacts from Address Book to a table whose header row contains the field names you want.

3. **In Address Book, select the contacts you want to add, and then click and drag them to the table.** Numbers automatically inserts the data in the appropriate columns and expands the table if necessary to accommodate the data (see figure 6.13).

Work Phone					
	A	B	C	D	E
1		First Name	Last Name	Work Phone	
2		Gina	Aaronovitch		
3		Nichola	Abbot	707-829-8301	
4		John	Abrahams		
5		Maria	Acton	212-309-8392	
6		Ron	Adinov	303-582-8191	
7		Craig	Anders		
8		Chris	Buvda		
9		Gerald	Dixon	510-555-1838	
10		Art	Eldrich		
11		Holly	Gehleicht		
12		Gary	Height	(707) 555-2314	
13		Jack	Maurer		
14		Jonno	Schmidt		
15					

6.13 Numbers automatically extracts the data from the contact records and snaps it into place.

Genius

You can also drag vCard files to a table in Numbers. A vCard file is a virtual business card that contains contact information in a computer-friendly format. So if you receive a vCard attached to an email message, as often happens, you can drag it directly to Numbers to add that person's details to a spreadsheet.

How Do I Perform Calculations in Numbers Spreadsheets?

Performing calculations is a vital part of most spreadsheets, and Numbers puts phenomenal power at your fingertips with built-in functions you can insert in moments. When these aren't enough, you can build your own formulas that perform custom calculations, adjust them by telling Numbers the order in which to evaluate them, and use the Formula List window to keep tabs on what's where. When you build a table, you can save it for reuse; and when you create an entire spreadsheet document that you want to use again, you can turn it into a template and use it directly from the Template Chooser.

Understanding Formulas

To perform a calculation in Numbers, you enter a formula in a cell. The *formula* gives the instructions for the calculation you want to perform. The cell containing the formula is called the *formula cell*.

Instead of displaying the text of the formula in the formula cell, Numbers displays the formula's *result*, the number or other value it produces. For example, if you enter the formula =SUM(B3:B5) in cell B6, telling Numbers to add the values in the range B3:B5, Numbers displays the result in cell B6.

If you then change any of the values in the range B3:B5, Numbers updates the formula result in cell B6 to show the new total. And if cell B6 appears in any other formulas, Numbers updates those too — and any formulas that use the results of those formulas, and so on until the whole spreadsheet is up to date with the latest figures.

Numbers formulas use three components: values, operators, and functions. Here's what these terms mean in Numbers:

- **Values.** A *value* can be text or a number that you type into a formula, but normally it's text or a number that you enter in a table cell and then refer to in the formula by using a cell reference. When you enter the data in a table cell, you can change it without having to edit the formula.

Note
A value that you enter directly in a formula is called a *constant* because it remains the same unless you change the formula.

- **Operators.** An *operator* is a symbol you use to tell Numbers which operation to perform on the value or values in a formula. For example, in the formula =B3/B2, the forward slash (/) is the division operator and tells Numbers to divide the value in cell B3 by the value in cell B2. The equal sign at the beginning tells Numbers that you're creating a formula rather than typing text.

- **Functions.** A *function* is a predefined formula that you can use in your tables. Most functions need one or more *arguments*, pieces of data on which to operate; a few, such as the NOW() function (which inserts the date and time) and the RAND() function (which inserts a random number), need no arguments. For example, when you use the SUM() function, you need to tell it which numbers or cell references to add. To do so, you enter the numbers, cells, or range as the argument — for example, =SUM(1,2,3) or =SUM(B3:B5).

Genius A Numbers function is like a black box: You can't dig inside the function and change what it does. If Numbers doesn't provide a function for the calculation you want to perform, you can write your own formula instead.

Inserting Functions

With Numbers, you can insert functions in your tables quickly and easily using the common calculations pane, the Function pop-up menu on the toolbar, or the Function Browser.

Inserting a function from the common calculations pane

The common calculations pane in the lower-left corner of the Numbers window enables you to quickly insert any of five widely used functions:

- **SUM().** Adds the specified values.

- **AVG().** Shows the average of the specified values.

- **MIN().** Shows the smallest of the specified values.

- **MAX().** Shows the largest of the specified values.

- **COUNT().** Shows the number of values.

Here's how to insert a function using the common calculations pane:

1. **Select the cells that contain the values you're interested in.** The buttons in the common calculations pane display the results of the formulas when applied to those cells.

2. **Click the function you need and drag it to the cell in which you want to place it.** Figure 7.1 shows an example of dragging the SUM() function to a cell.

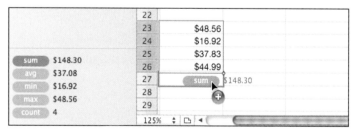

7.1 The quickest way to insert a function is by dragging one of the five functions from the common calculations pane to a cell.

Even if you don't insert any of the functions from it, the common calculations pane is also handy for reference. For example, to see how many cells you've selected, look at the readout next to the Count button.

Inserting a common function from the toolbar

The second way to enter functions is to use the Function pop-up menu on the toolbar. As you can see in figure 7.2, this pop-up menu offers the same five functions as the common calculations pane, plus the PRODUCT() function (which multiplies the values you specify), and commands for displaying the Function Browser and Formula Editor (both of which you'll meet shortly).

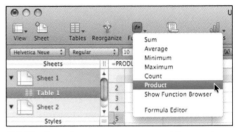

7.2 The Function pop-up menu enables you to insert six widely used functions and open the Function Browser and Formula Editor.

Note

You can also insert functions from the Insert ⇨ Function submenu on the menu bar. Usually the only reason to use this submenu rather than the Function pop-up menu is if you've hidden the toolbar to get more space onscreen.

When you use the Function pop-up menu to insert a function, you can select the value cells either beforehand or afterward. When you select the value cells first, they must all be either in the same row or in the same column. Numbers can then automatically choose the formula cell, as follows:

- **Column.** If the value cells run down a column, Numbers places the formula cell below the bottom value cell.

- **Row.** If the value cells run across a row, Numbers places the formula cell to the right of the rightmost value cell.

This behavior works well when you want the formula cell in these locations, but other times you'll want to place it elsewhere. In this case, place the formula and then choose the value cells like this:

1. **Click the cell in which you want to place the formula.**

2. **Click the Function pop-up menu on the toolbar, and then click the function you want.**
   Numbers inserts the function in the formula cell and selects the range of cells it guesses you want to use for the calculation — for example the cells above the formula cell, as in figure 7.3. Numbers displays the Formula bar buttons in place of the Format bar.

=SUM(E2:E5)

	A	B	C	D	E	F
1	State	Ohio	Arizona	Idaho		
2	2009	49	14	19		
3	2010	18	18	37		
4	2011	63	73	25		
5	2012	24	61	43		
6					0	
7						

7.3 When there's no obvious range of cells to use for a function you enter, Numbers suggests the range of cells above or to the left of the formula cell.

3. **Choose the value cells you want to use.** If Numbers has chosen the right range, simply accept its suggestion. Otherwise, click the range shown in the Formula bar, and then click and drag in the table to select the cells you want. The mouse pointer shows a function symbol, as shown in figure 7.4.

=SUM(B5:D5)

	A	B	C	D	E	F
1	State	Ohio	Arizona	Idaho		
2	2009	49	14	19		
3	2010	18	18	37		
4	2011	63	73	25		
5	2012	24	61	43		
6					0	
7						

7.4 You can quickly select the range of cells you want instead of Numbers' suggestion.

Note If you select the wrong range of value cells, simply press Esc to cancel the changes you're making.

4. **Press Return to apply the change to the formula.** Numbers shows the formula's new result and replaces the Formula bar buttons with the Format bar.

Inserting a function with the Function Browser

The handful of functions that the common calculations pane and the Function pop-up menu provide are widely useful, but often you'll need to use some of Numbers' other 250-odd functions. Your main tool for doing so is the Function Browser.

Here's how to insert a function with the Function Browser:

1. **Click the cell in which you want to place the function.**

2. **Open the Function pop-up menu on the toolbar and choose Show Function Browser.** Numbers displays the Function Browser (see figure 7.5).

7.5 The Function Browser lets you insert any of Numbers' many functions.

Genius

If you need to see more information about the functions at once, you can expand the Function Browser by dragging the handle in its lower-right corner.

3. **In the list box in the upper-left corner, click the category of function you want.** For example, click Date and Time if you want to see functions for calculating dates or times. You can also choose All or Recent, or search for a function:

   ● **All.** This category shows you all the functions. It's a bit unwieldy, but it's handy if you're not sure which category the function belongs to.

Recent. This category shows you functions you've recently used. If you use the same functions frequently, looking here can save you time.

 Search. To search for a function, type the name or keyword in the search box at the top of the Function Browser. The Function Browser displays a list of all functions that have that term in their name or explanations.

4. **When you've found the function you want, click it, and then click Insert Function.** Numbers inserts the function in the formula cell and displays the Formula Editor (see figure 7.6).

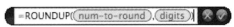

7.6 In the Formula Editor, enter the cell references for the function, and then click the Accept button (the green check mark).

Note

If you've finished with the Function Browser for the time being, click its Close button to close it. Otherwise, Numbers keeps the Function Browser open in case you need it.

5. **Enter the cell references for the function in the Formula Editor. Click an argument, and then click the cell you want to use.** You can enter a constant by clicking an argument in the Formula Editor and then typing the value over it.

6. **When you've entered all the arguments, click the Accept button (the green check mark) or simply press Return to close the Formula Editor.** Numbers displays the result of the function in the cell.

Note

You can display the Formula Editor at any time by double-clicking a cell that contains a formula, or by clicking the cell to select it and then clicking it again (slower than a double-click). Alternatively, use the arrow keys to select the formula cell and press Option+Enter to display the Formula Editor.

Typing a function into a cell

Instead of using the common calculations pane, the Function pop-up menu, or the Function Browser, you can simply type a function into a cell. As soon as you start typing, Numbers opens the Formula Editor for you.

When you know the name of the function you need, typing it in is usually quicker than picking the function from the Function Browser.

Dealing with errors in functions

If Numbers displays an orange triangle in a cell in which you've inserted a function, there's a problem with the function — for example, you haven't told the function which arguments to use, or you've asked it to do something impossible, such as dividing by zero.

Click the cell with the orange triangle to pop up a message explaining the problem (see figure 7.7). You can then open the Formula Editor and fix the problem — for example, by supplying any missing arguments or by locating the cell that is providing the zero value by which the function is trying to divide.

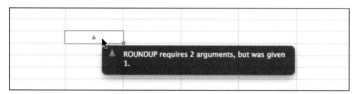

7.7 Click a cell bearing an error marker to see what the problem is.

Finding Information about Numbers' Functions

You can find out what a function in Numbers does three main ways:

- **Use the Function Browser information.** Open the Function Browser, select the function, and look at the information provided. If you need to see more at once, click and drag the sizing handle in the lower-right corner of the Function Browser down, to the right, or both.

- **Read the *iWork Formulas and Functions User Guide*.** You can download this PDF file from the Apple Web site (http://manuals.info.apple.com/en_US/Formulas_and_Functions_User_Guide.pdf). It's a hefty read, but it's a great way to get an overview of all the functions Numbers offers, plus a fair amount of detail on what they do.

- **Read the Help files.** From Numbers, choose Help⇨iWork '09 Formulas and Functions Help. You can then view the Overview of the iWork Functions topic, the Function Reference topic, or the topic called Additional Examples and Topics.

To find out everything you need to know to use the functions effectively, you may need to combine these different ways of finding information about the functions. At this writing, the *iWork Formulas and Functions User Guide* for iWork '09 is not as complete as the *iWork Formulas and Functions User Guide* for iWork '08 — but some of the functions have been updated for Numbers '09, so you may do better to look them up in the Function Browser or Help files instead.

Creating Your Own Formulas

Numbers' built-in functions are great for performing the wide variety of tasks for which they're designed. But unless you create unusually simple spreadsheets, you'll also need to create formulas of your own that perform exactly the calculations you need.

You can create a formula either in the Formula Editor or in the Formula bar. The process is much the same in either tool, and you can move from one to the other as needed. You'll probably want to start with the Formula Editor, as described in a moment.

- **Formula Editor.** The Formula Editor is a pop-up window that appears next to the cell in which you're creating the formula. The advantage of the Formula Editor over the Formula bar is that it's near to the data you're using, and you can reposition it wherever you want within the Numbers window.

- **Formula bar.** The Formula bar's editing box appears under the Format bar, and its buttons appear in place of the Format bar when you start creating a formula. You can click the handle at the right end of the Formula bar and drag it downward to give yourself more space for the formula. You can also change the size of the text used in the formula by using the Formula Text Size pop-up menu. This is handy if you find the text size in the Formula Editor too small to read comfortably.

Creating a formula with the Formula Editor

Here's how to create a formula with the Formula Editor:

1. **Open the Formula Editor:**

    - The easiest way is to click the cell and then press =, but you can also click the cell, open the Function pop-up menu, and then choose Formula Editor.

    - If the cell already contains a formula, you can open the formula for editing in the Formula Editor by double-clicking or by pressing Option+Return.

Note

If the Formula Editor opens over a cell you need to see, click the curved area at the left end and drag the Formula Editor to somewhere more convenient.

2. **Build your formula by doing the following in whichever combination you need:**

    - **Add a cell reference.** Click the cell you want to add. Numbers adds it to the Formula Editor as a rounded rectangle that shows the cell reference.

Note If the table has header columns and header rows, the cell reference normally appears as a combination of the names. For example, if the column has the header Arizona and the row has the header 2012, the cell reference appears as Arizona 2012. If the table has no headers, the cell reference appears as the combination of the column letter and row number — for example, C3.

- **Insert an operator.** Click where you want to place the operator, and then type it. You can skip typing the addition operator (+) because Numbers automatically inserts the addition operator when you click a series of cells in succession. Numbers uses a different color for each cell reference so that you can instantly see which reference in the Formula Editor maps to which cell. Figure 7.8 shows an example.

	A	B	C	D	E	F
	fx Cancel Accept	Formula Text Size: 10				
	=(Idaho 2012)+(Arizona 2012)+(Ohio 2012)+(Ohio 2010)					
1	State	Ohio	Arizona	Idaho		
2	2009	49	14	19		
3	2010	18	18	37		
4	2011	63	73	25		
5	2012	24	61	43		
6						
7		=(Idaho 2012)+(Arizona 2012)+(Ohio 2012)+(Ohio 2010)				
8						
9						

7.8 Numbers automatically uses different colors for cell references in the Formula Editor and the Formula bar so that you can easily tell which cell references go where.

- **Insert a function.** In the Formula Editor, click where you want the function to appear, then either click the Function Browser (fx) button in the Formula bar or open the Function pop-up menu on the toolbar and choose Show Function Browser. Choose the function as described earlier in this chapter, and then click Insert Function to insert it. Add any arguments the function needs.

- **Delete an element from the formula.** Click the element, and then press Delete.

- **Rearrange the elements in the formula.** Click and drag the element to its new position.

- **Insert a line break or tab.** To make a long formula easy to read, you can break it onto separate lines or insert tabs to indent parts of it. Click the Line Break button (the button with a Return-key symbol) at the right end of the Formula bar to insert a line break; click the Tab button (which bears a tab symbol and is next to the Line Break button) to insert a tab.

Any line breaks or tabs you insert in your formulas are simply for your benefit. Numbers ignores them, so they don't affect the result of the formula.

3. **When you've finished creating the formula, click Accept in the Formula bar or press Return to enter it.** If you decide you don't want to keep the formula, click Cancel or press Esc instead.

Creating a formula in the Formula bar

To create a formula in the Formula bar, simply click in the cell, and then click in the editing area of the Formula bar. Numbers hides the Format bar and displays the Formula bar's controls in its place.

You can then type an equal sign (=) to start creating the formula. Continue by using the techniques described in the previous section, and click Accept in the Formula bar (or press Return) when you've finished.

Referring to cells

To make your formulas work correctly, you need to tell Numbers which cells to use. You do this by using cell references that identify the cells uniquely in the spreadsheet document.

You can refer to cells two ways:

- **Using A1-style references.** As in most spreadsheet applications, you can use the letter of the column reference tab and the number of the row reference tab that intersect at the cell. For example, the cell at the intersection of column A and row 1 is cell A1.

- **Using header names.** You can use the name of the column header and the name of the row header. For example, to refer to the cell at the intersection of the column with the header Arizona and the row with the header 2012, you can use Arizona 2012 (with a space between the names).

If you prefer "old-fashioned" cell references such as A1, B2, and C3 rather than references that use header names, open General Preferences (choose Numbers ⇨ Preferences) and deselect the Use header cell names as references check box. You can then click the cells you want to use in your formulas, and Numbers will use A1-style naming for the references.

As you'd expect, you can use header names only when you've added header columns and header rows to a table. Until then, you must use A1-style references.

Entering references quickly using the Option key

When creating a formula using the keyboard, you can quickly enter references by using the Option key. Here are the details:

- **Enter a single cell reference.** Hold down Option and use the arrow keys to select the cell.

- **Enter a reference to a range.** Hold down Option, select the first cell, and then hold down Shift+Option while you select the rest of the range.

- **Move to another table.** Press ⌘+Option+Page Down to move to the next table or ⌘+Option+Page Up to move to the previous table.

Referring to another table on the same sheet

To refer to a cell in another table on the same sheet, click the table in the Sheets pane to display it, and then click the cell. Numbers automatically inserts the correct reference for you — but you may need to adjust it.

If the cell reference derived from the column tab (or header) and row tab (or header) of the cell you click is unique in the spreadsheet document, Numbers creates the reference the same way as for a reference within the table. For example, if only one table in the spreadsheet document has a cell named Ohio 2011 (from a column headed Ohio and a row headed 2011), Numbers enters the reference Ohio 2011.

But if any other table on the sheet, or elsewhere in the spreadsheet, contains a cell with the same name, Numbers creates the reference with the table's name first, a space, then two colons, another space, and then the cell reference. For example, if the Ohio 2011 cell is in the States table, the reference is States :: Ohio 2011.

Caution

Using the cell reference without the table name (or the sheet name, as discussed next) can lead to mistakes if you subsequently create another table that reuses header names already used by other tables. To avoid ambiguity, add the table name (and if necessary the sheet name) even if Numbers doesn't. Double-click the cell reference to open it for editing, type in the extra information, and then press Return or click the Accept button.

Referring to a table on another sheet

Referring to a cell in a table on another sheet works the same way referring to another table on the same sheet does, except that you may need to add the sheet's name as well as the table's.

To insert the reference, click the sheet's disclosure triangle in the Sheets pane if it's collapsed, click the table, and then click the cell. Numbers inserts the reference for you.

Again, the cell reference Numbers inserts varies depending on the column, row, and table names:

- If the cell reference is unique in the spreadsheet, Numbers inserts the reference without the table name or sheet name.

- If the cell reference is not unique but the table name is, Numbers inserts the reference with the table name but without the sheet name.

- If neither the cell reference nor the table name is unique, Numbers inserts the sheet name as well, again using a space, two colons, and another space as separator characters.

For example, if the Ohio 2011 cell is in the States table on the Delegates sheet, the full reference, including the sheet name and the table name, is Delegates :: States :: Ohio 2011.

If Numbers doesn't insert the full reference, you may want to do so yourself to prevent confusion if you add sheets or tables.

Choosing between absolute and relative references

Like most other spreadsheet applications, Numbers lets you use absolute references, relative references, and mixed references.

- **Absolute reference.** This is a reference that always refers to the same cell, even if you move the formula to a different formula cell. For example, say you enter 2 in cell B2 and 4 in cell B3. If you enter the formula =$B$2+$B$3 in cell B4, it always adds the values in cell B2 and cell B3, even if you move the formula from cell B4.

Note

Numbers uses the dollar sign ($) to indicate that part of a reference is absolute. For example, the reference $A$1 always refers to cell A1, and the reference $Arizona $2010 always refers to the cell in the column with the header Arizona and the row with the header 2010.

- **Relative reference.** This is a reference that is relative to the formula cell's position in the table. For example, if you enter the formula =B2+B3 in cell B4, it adds the values in cell B2 and cell B3, but the underlying meaning of the formula is "add the cell two cells above the formula cell to the cell directly above the formula cell." So if you move the formula to cell C4, the formula adds the values in cell C2 and cell C3 instead. This behavior is handy when you need to copy formulas or rearrange tables.

- **Mixed reference.** This is a reference that is absolute for either the column or the row, but not for both. For example, if you enter the formula =B$2+B$3 in cell B4, it tells Numbers to add the second cell and third cell (the row reference is absolute) in the same column as the formula cell (the column reference is relative). If you move the formula to cell C5, the formula still operates on rows 2 and 3, because that part of the formula is absolute, but on column C instead of column B.

When you click a cell in a table to add it to a formula, Numbers automatically creates a relative reference. You can then change it to an absolute reference or mixed reference.

The easiest way to change the reference type is to highlight the reference in the Formula Editor or the Formula bar, click the disclosure triangle, and then choose the reference type from the pop-up menu (see figure 7.9).

You can also click a cell reference in the Formula Editor or Formula bar and then press ⌘+K to cycle through the four reference types:

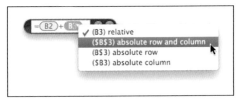

7.9 Changing the reference type using the pop-up menu in the Formula Editor.

- **Absolute row and column.** For example, $B$3.

- **Absolute row, relative column.** For example, B$3.

- **Relative row, absolute column.** For example, $B3.

- **Relative row and column.** For example, B3.

Note You can also type an absolute reference or mixed reference using the keyboard. Simply place a dollar sign before each part of the reference you want to make absolute.

Copying or moving a formula

You can copy or move a formula cell just as you can copy or move the content of any other cell. For example, select a formula cell, press ⌘+C to copy its contents or ⌘+X to cut its contents, move to the destination cell, and then press ⌘+V to paste in the formula.

When you paste a formula with a relative component, Numbers changes the cells the formula refers to so that they match the formula cell's new location. For example, if you have a formula in cell B4 that refers to cell B3, moving the formula to cell C4 makes it refer to cell B3 instead.

By contrast, any absolute reference still refers to the same cell no matter where you place it. A mixed reference still refers to its absolute column or absolute row, but its relative component changes to reflect its new location.

Deleting a formula

To delete a formula, click its cell and then press Delete.

If you want to see the formula in context with the other formulas in the spreadsheet, open the Formula List window (discussed later in this chapter). You can then click a formula in the Formula List and press Delete to delete it.

Understanding Operators and How Numbers Evaluates Them

To perform arithmetic or comparisons in your formulas, you use operators. For example, to add values, you use the addition operator, the plus sign (+); to see if one value is larger than another, you use the greater-than operator (>). To join together text items, you use the string operator.

Arithmetic operators

Table 7.1 shows the six arithmetic operators that Numbers uses. They're all straightforward to use, but there are two things to watch out for:

- **Text.** If you try to perform arithmetic on text, you get an error. Numbers alerts you that you've given it a "string" rather than a number. (A *string* is nonnumeric text — for example, your name.)

- **Division by zero.** Dividing by zero is mathematically impossible, so Numbers gives an error when a formula tries to do so. If the cell in question doesn't contain zero or a calculation that produces zero, it may be blank (which Numbers treats as zero).

Table 7.1 Numbers' Arithmetic Operators

Arithmetic Operation	Operator
Addition	+
Subtraction	–
Multiplication	*
Division	/
Exponentiation	^
Percentage	%

Comparison operators

Numbers has six comparison operators you can use to compare values. Table 7.2 shows them. There are four main things to note here:

- **Return values.** Each comparison operator returns TRUE if the condition is true and FALSE if it is not true.

- **Strings are "greater than" numbers.** Numbers considers any string to be "greater than" any number. So, for example, the formula **="cat">10000000** returns TRUE. You probably won't enter nonsensical comparisons like this in your spreadsheets, but if you're getting an unexpected TRUE or FALSE value, look to see if a string is causing it.

- **Numeric values of TRUE and FALSE.** Numbers assigns the value 1 to TRUE and the value 0 to FALSE. That means TRUE > FALSE. (Most social scientists think otherwise.)

- **TRUE and FALSE compare only with each other.** Despite TRUE and FALSE having the values 1 and 0, you can't compare them with numbers. For example, using TRUE = 1 returns FALSE even if whatever you're checking is true. You also can't compare TRUE or FALSE with strings of text; doing so always returns FALSE.

Table 7.2 Numbers' Comparison Operators

Comparison Operation	Operator
Equal	=
Not equal	<>
Greater than	>
Smaller than	<
Greater than or equal to	>=
Smaller than or equal to	<=

The string operator and the wild cards

Apart from the arithmetic operators and comparison operators, Numbers uses one string operator and three wild card characters. Table 7.3 shows these operators.

Table 7.3 Numbers' String Operator and Wildcards

Operation	Operator or Wildcard	Explanation
Join strings or cell contents	&	Joins the strings of text or the values of cells, treating them as text even if they're numbers.
Match one character	?	Use this to match any one character. For example, "mi?" matches any three-character string beginning with "mi."
Match multiple characters	*	Use this to match any number of characters. For example, "dr*" matches "dry," "drive," and many other strings.
Match a wildcard	~	Use "~?" to match a real question mark instead of using the question mark as a wild card.

Note

The string operator can give you surprise results in your calculations. For example, the formula "=50&100" returns 50100 rather than 150. While you're not likely to enter the string operator directly in a formula like that, it can creep in when you're creating a formula with cell references.

Overriding the order in which Numbers evaluates operators

When you put together a formula that contains two or more operators, you sometimes need to know the order in which Numbers will evaluate them.

For example, if someone offers to give you the result of 2000–200*5 dollars, will you get $9,000 (2000 – 200 =1800; 1800 × 5 = 9000) or only $1,000 (200 × 5 = 1000; 2000 – 1000 = 1000)?

Sadly, you would get only $1,000, because Numbers performs the multiplication before the subtraction. Table 7.4 shows the order in which Numbers evaluates operators, from first to last.

Table 7.4 Operator Precedence in Numbers (Descending Order)

Operator	Explanation
%	Percentage
^	Exponentiation
* and /	Multiplication and division
+ and −	Addition and subtraction
&	Concatenation
=, <>, <, <=, >, >=	Comparison operators

You can override this order by putting parentheses around the item you want to evaluate first in the formula. For example, (2000-200)*5 makes Numbers evaluate 2000–200 first, and then multiply the result of that by 5, giving 9000.

Genius

You can put one item in parentheses inside another item in parentheses if necessary. This is called *nesting* items, and Numbers performs the deepest nested calculation first. For example, in (2000-(200-100))*5, Numbers calculates 200-100 first, so the formula gives a result of 9500. As you enter each closing parenthesis, Numbers highlights its matching opening parenthesis to help you keep track of which calculation you're closing.

Checking Your Formulas with the Formula List

When you build a spreadsheet stuffed with formulas, it's easy to lose track of exactly which formula is where. Numbers' Formula List window lets you easily come to grips with all the formulas in a document.

Here's how to open the Formula List window:

- **Keyboard.** Press ⌘+Option+F.
- **Mouse.** Choose Show Formula List from the View pop-up menu on the toolbar or the View menu on the menu bar.

The Formula List window opens at the bottom of the Numbers window (see figure 7.10). You can increase the depth of the Formula List window by dragging the sizing handle to the right of the Search box upward (or decrease it by dragging the handle downward).

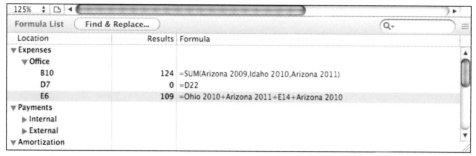

7.10 Use the Formula List window to quickly check the formulas contained in the current spreadsheet document.

You can use the Formula List window to find formulas three ways:

- **Expand or collapse the listing.** Click a disclosure triangle to hide the tables in a sheet or the formulas in a table; click again to display them once more.

- **Search for a formula.** Type the search term in the Search box. To reuse a recent search, click the drop-down arrow on the Search icon, and then click the search term.

- **Use Find & Replace.** Click the Find & Replace button to open the Find & Replace dialog box. You can then search for formula items or replace them wholesale.

Creating Your Own Reusable Tables

In many jobs, you'll need to create similar types of tables in the same spreadsheet. You can copy a table from one sheet and paste it onto another sheet, but Numbers also provides an easier way of keeping the tables you want at hand: You can "capture" a table for future use so that you can quickly insert it elsewhere in the document.

Note Saving a table as described here makes the table available only in the Numbers document you're using. To make a table available in other Numbers documents, you need to save a custom template, as described later in this chapter.

Capturing a table

When you've created and formatted a table you want to keep, capture it like this:

1. **Choose Format ⇨ Advanced ⇨ Capture Table.** Numbers displays the New table dialog box (see figure 7.11).

2. **Type a descriptive name for the table in the Name box.** You can call the table pretty much whatever you want as long as there's no predefined table with that name. (If there is, Numbers asks whether you want to replace it when you click OK.)

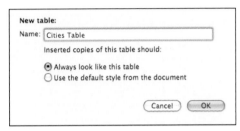

7.11 Use the New table dialog box to capture a table so that you can quickly insert it on other sheets in the same document.

3. **Choose the style for the table from the Inserted copies of this table should options:**

   - **Always look like this table.** Select this option button if you want the table to have the formatting you've applied. This is usually the best choice if you've formatted the table to suit the template.

   - **Use the default style from the document.** Select this option button if you want just the table's content. The table takes on the default style of the Numbers template. As usual, you can reformat the table if you want.

4. **Click OK.** Numbers saves the table in the document.

Inserting a captured table

After you capture a table, you can insert it on any sheet in the document by opening the Tables pop-up menu on the toolbar and choosing the table from the menu.

Managing your custom tables

When you've added custom tables to a spreadsheet document, you can use the Manage Table Prototypes dialog box to change the order in which the tables appear on the Table pop-up menu. You can also rename tables or delete ones you no longer need.

Here's how to manage your custom tables:

1. **Open a spreadsheet that contains the custom tables.**

2. **Choose Format ➪ Advanced ➪ Manage Tables.** Numbers opens the Manage Table Prototypes dialog box (see figure 7.12).

3. **Click the table you want to affect, and then take one of these actions:**

   ● **Move the table up or down the list.** Click the Up button or the Down button.

   ● **Rename the table.** Double-click the table's name, type the new name, and then press Return.

   ● **Delete the table.** Click the Delete button.

4. **Click Done to close the Manage Table Prototypes dialog box.**

7.12 Use the Manage Table Prototypes dialog box to change the order of tables, rename tables, and to delete any tables you don't need.

Creating a Template from a Spreadsheet

When you've created a spreadsheet that contains sheets, tables, or charts you want to reuse easily, save the spreadsheet as a template, as discussed in Chapter 1. You can then select the template from the Template Chooser when you want to create a new spreadsheet based on the template.

Before you save a spreadsheet as a template, set it up as fully as possible. Perform as many of these actions as your spreadsheet needs:

● **Remove any unnecessary sheets or tables.** This should go without saying, but it's amazing how many templates you'll find that have rough sheets or draft versions of tables that the creator forgot to remove.

● **Strip out the variable data from the tables.** If you've created a final spreadsheet, the tables are probably full of the data that this version of the spreadsheet needs — but that other spreadsheets based on your new template will not need. Remove all the variable data to avoid any confusion about whether it belongs there.

Caution

When creating a template, be careful not to include sample data that can be mistaken for real data. Instead, insert clear instructions in cells where they're needed, and attach comments to any cells that need additional explanation.

- **Add comments to anything ambiguous.** The spreadsheet may be clear as glass to you because you've created it, but it's likely to be opaque to anyone else. If you've ever puzzled over how someone else's template is supposed to work, or which data goes where, you'll appreciate the value of helpful, concise comments.

- **Create any table styles needed.** If you've applied custom formatting to the tables in the spreadsheet, make sure you turn them into styles so that they're easy to apply to new tables that users of the template create. See Chapter 8 for instructions on creating table styles.

- **Capture any reusable tables needed (as discussed earlier in this chapter).** If users of the template will need to insert tables, make it easy for them by providing the tables right on the Tables pop-up menu.

- **Set up any default charts (as discussed in the next chapter).** If users of the template will need to insert charts, you can make this easy for them as well.

How Can I Make My Spreadsheets Dynamic?

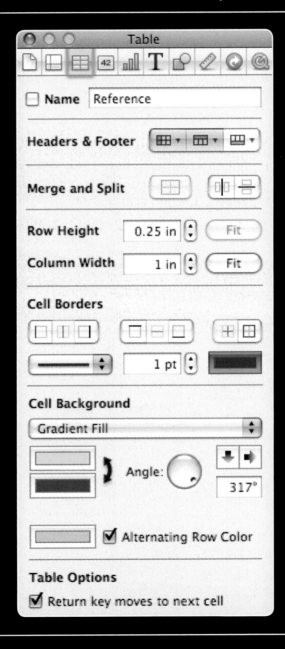

When you've entered all the data the spreadsheet needs, it's time to make the spreadsheet look compelling. Start by using Numbers' table styles to format each table, using conditional formatting to monitor cells for unexpected values, and sorting and filtering each table so that it shows only the data you want to see. You can create powerful charts that draw data from different tables, share them with Pages and Keynote, and even update them from within a document or a presentation. For extra visual impact, you can add images to a sheet, cell, table, or other object; and for convenience, you can add dynamic controls that let the user manipulate values accurately and easily.

Making a Table Look Exactly How You Want It

This section shows you how to make each table look exactly the way you want it by applying table styles, using conditional formatting to automatically monitor cells for unexpected values, and filtering a table so that it shows only the data you're interested in.

Formatting a table with a table style

A quick way to make a table look good is to apply a table style from the Styles pane. A *table style* is a complete set of formatting for a table, including everything from the text formatting for body cells to the border formatting for header and footer cells, plus any background image or color the table as a whole needs.

Each table always has a style applied to it, even if it's only the Basic style that makes the table appear as a grid with no formatting. In many of Numbers' templates, each table comes with one of the colorful styles applied to give the table its look. To see which table style is applied to a table, click anywhere in the table, and then look at the Styles pane to see which style is highlighted.

To apply a different style, just click anywhere in the table — you don't need to select the whole table — and then click the style in the Styles pane. Numbers applies the style to the entire table. You can also click and drag a style from the Styles pane to any table on the sheet canvas, which is handy for formatting a table that you haven't selected.

Genius

While you can't apply a style to only part of a table, you can create different tables, apply different styles, and place the tables next to each other to give the appearance of using different styles in different parts of the table. If you remove the header columns and rows and footer rows where the tables meet, the tables appear to make up a single table.

If you've applied any formatting to the table, Numbers keeps that formatting when you apply the new style. For example, if you've applied bold or italics to a cell, that formatting will still be there after you change the style.

Usually, this feature is helpful, but sometimes you may want to remove any formatting you've applied so that the whole table has that style. To do this, click the table, highlight the style in the Styles pane, click the pop-up button, and choose Clear and Apply Style.

Formatting table cells and borders manually

The table style gives you the broad strokes of the table's formatting, but you'll often need to format some table cells manually to make them stand out.

You can quickly change the border formatting or fill color of one or more selected cells by using the border and fill controls on the Format bar (see figure 8.1). From the Borders pop-up menu, choose the borders you want to change, or click Allow Border Selection, click the borders on the table, and then click Disallow Border Selection. You can then choose the line style and line width for the selected borders.

For greater control of borders, or to use a gradient fill or image as the background for the cell, open the Table Inspector by clicking the Inspector button on the toolbar and then clicking the Table Inspector button. You can then use the controls in the Cell Borders area and the Cell Background area to change the borders and background of the cells.

8.1 Use the Format bar's border and fill controls to quickly change borders and background color.

Monitoring cells for unexpected values

If your spreadsheet contains important data, set conditional formatting on vital cells so that Numbers automatically alerts you when a cell contains an unexpected value. For example, in an expense report, you can make Numbers automatically highlight any item greater than the amount you set. Figure 8.2 shows an example.

Establishment	Location	Cost
Staples	Emeryville	$99.43
Greasy Burger Emporium	Oakland	**$100.00**
Best Buy	Emeryville	$14.99
Total Due:		$214.42

8.2 Use conditional formatting to automatically pick out unexpected values in a table. Here, the Cost column has a threshold of $100.

Here's how to apply conditional formatting to cells:

1. **Select the cell or cells you want to affect.** If you need to monitor an entire row or column that contains the same type of data, select that row or column by clicking its reference tab.

2. **Open the Cells Inspector.** Click the Inspector button on the toolbar, and then click the Cells Inspector button.

3. **In the Conditional Format area, click the Show Rules button.** Numbers displays the Conditional Format dialog box. At first, this dialog box appears as shown in figure 8.3, with just the Choose a rule item selected in the pop-up menu.

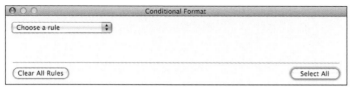

8.3 Start applying conditional formatting by opening the Choose a rule pop-up menu.

4. **Open the Choose a rule pop-up menu and choose the criterion for the first (or only) rule.** Your choices are: Equal to, Not equal to, Greater than, Less than, Greater than or equal to, Less than or equal to, Between, Not between, Text contains, Text doesn't contain, Text starts with, Text ends with, Text is, and With dates. This example uses the Greater than or equal to criterion.

5. **Enter the value or values in the field or fields that the Conditional Format dialog box displays.** Either simply type a value in the field, or click the blue button at the right end of the field to tell Numbers you're using a cell reference, and then click the cell in the table.

Genius

Tie your conditional formatting to a value in a cell rather than typing the value directly into the Conditional Format dialog box. You can then quickly change the trigger value in the cell rather than having to open the Conditional Format dialog box to change it. To keep your tables tidy, set up a separate table that contains the values for all the conditional formatting in the spreadsheet document.

6. **Change the formatting that the condition will apply.** The Sample cell in the Conditional Format dialog box shows how the cell will appear when the cell's contents meets the condition. To change it, click the Edit button and use the line of buttons that the Conditional Format dialog box displays to set the formatting (see figure 8.4). For example, click the Fill color box and choose a fill color that will stand out from the rest of the table formatting, or click the Bold button to apply boldface. Click Done when you've finished.

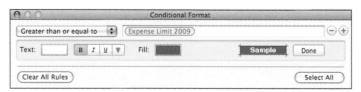

8.4 Use the formatting controls to set up the formatting you want Numbers to apply when the condition is met.

7. **If necessary, click the + button at the right end of the first rule to add another rule.** Create the rule and specify formatting for it using the techniques explained in Steps 4 through 6.

8. **When you've finished setting up conditional formatting, click the red button in the upper-left corner to close the Conditional Format dialog box.**

When you need to remove conditional formatting from one or more cells, open the Conditional Format dialog box again. You can then click the Select All button to see which cells have a particular type of conditional formatting applied to them. To remove a single rule, click the – button at its right end. To remove all rules, click the Clear All Rules button.

Genius

Conditional formatting can take some time and effort to get right, but you can save time by copying conditional formatting you've created for one table to another table.

Sorting and filtering a table to show results

To make a table show data in the most useful way, you may need to sort it or filter it:

- **Sorting.** Sorting lets you rearrange the rows into a different order. For example, if you have a table of sales by rep, you can sort by the sales total column so that you can see which rep sold the most.

- **Filtering.** Filtering lets you narrow down a table so that it shows only the rows you're interested in. For example, you can filter a table of customers so that it shows only those with addresses in Buffalo.

Sorting a table

To sort a table quickly, highlight the header column by which you want to sort, click the pop-up button, and then choose Sort Ascending or Sort Descending. Numbers sorts the entire table; you can't sort just part of the table by selecting it first.

Note

An ascending sort sorts values from smallest to largest, from A to Z, from earlier dates or times to later ones, and puts numeric values before text values. A descending sort sorts values in the reverse order.

Here's how to sort only part of a table or to sort by multiple columns:

1. **Click anywhere in the table you want to sort.** If you want to sort only some rows, select them.

2. **Click the Reorganize button on the toolbar to open the Reorganize window.** The window's title bar shows the current table's name for reference — for example, Reorganize Sales.

3. **If the Sort section is hidden, click the Sort disclosure triangle to display it.**

4. **In the first pop-up menu, choose the header name or column letter by which to sort, and then choose ascending or descending in the second pop-up menu.**

5. **To add another sort column, click the + button at the right end of the first row of controls.** Choose the sort column and whether to use an ascending or descending sort.

6. **Repeat Step 5 to set up all the sort columns you need.** If you need to remove a sort column, click the – button on its row. If you need to change the order of sort columns, click the Up button or Down button on a row (see figure 8.5).

Up button

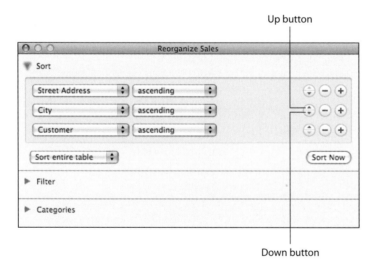

Down button

8.5 Use the Sort area of the Reorganize window to sort by two or more columns or to sort only some rows in a table.

7. **In the lower pop-up menu, choose Sort entire table or Sort entire rows, as needed.**

8. **Click Sort Now.** You can then either close the Reorganize window or change the sort order and sort again.

Filtering a table to show only some results

The Reorganize window also lets you make a table show only the values you want to see. By slimming down the table like this, you can focus clearly on your important data.

Here's how to filter a table:

1. **Click anywhere in the table you want to filter.**

2. **Click the Reorganize button on the toolbar to open the Reorganize window.**

3. **If the Filter section (the middle section) is hidden, click the Filter disclosure triangle to display it.**

4. **In the first pop-up menu, choose the first header name or column letter by which to filter.** The Reorganize window displays controls for setting the filtering criterion.

5. **In the second pop-up menu, choose the comparison.** Your choices are

   - "is" or "is not"
   - "is blank" or "is not blank"
   - "is greater than," "is less than," or "is between"
   - "is in the top," "is in the bottom," "is above average," or "is below average"
   - "contains," "doesn't contain," "starts with," or "ends with"
   - "date is"

6. **At the right end of the first row, use the controls that the Reorganize window displays to set the value for the comparison.** This depends on the comparison you chose. For example:

   - For a "date is" comparison, choose a specific date or a range of dates.
   - For an "is greater than" comparison, enter a value.
   - For an "is in the bottom" comparison, choose the percentage of values (for example, "is in the bottom 10 percent") or the number of items (for example, "is in the bottom 5 values").
   - for an "is blank" or "is not blank" comparison, you need choose nothing to compare beyond the blankness.

7. **Click the + button at the right end of the first row of controls.** Numbers reveals another line of controls, selects the Show rows that match check box, and applies the first filter to the table.

8. **If you want to filter by a second criterion, set it up on the second row using the techniques described in Steps 4 through 6.**

9. **Repeat Steps 7 and 8 to set up all the filtering columns you need.** Figure 8.6 shows the Reorganize window with filtering set for two columns.

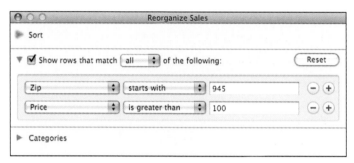

8.6 The Reorganize window also enables you to filter a table by one or more columns.

10. **Reorganize your filters as needed:**

   - To remove a filtering column, click the – button on its row.

   - To view rows that match any of your criteria rather than rows that match all of the criteria, choose Any in the Show rows that match pop-up menu.

   - To turn off filtering temporarily, deselect the Show rows that match check box.

   - To clear all your filters, click Reset.

11. **When you've finished filtering the table, close the Reorganize window.** You can either leave the filters in place (so that you see only matching rows), turn them off by deselecting the Show rows that match check box before closing the window, or get rid of them by clicking Reset.

Organizing a table by categories

To make large tables easier to read, you can organize rows into categories. A *category* is like a heading within the table; you can collapse the category to hide the rows that it contains. You can also create subcategories within any category, and collapse those as needed.

Figure 8.7 shows a table organized by Year categories and Month subcategories. The result is a collapsible table in which you can expand a year's data (by clicking its disclosure triangle) to reveal the month subcategories within it, and then expand a month subcategory to reveal its data.

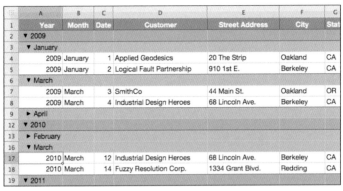

	A	B	C	D	E	F	G
1	Year	Month	Date	Customer	Street Address	City	Stat
2	▼ 2009						
3	▼ January						
4	2009	January	1	Applied Geodesics	20 The Strip	Oakland	CA
5	2009	January	2	Logical Fault Partnership	910 1st E.	Berkeley	CA
6	▼ March						
7	2009	March	3	SmithCo	44 Main St.	Oakland	OR
8	2009	March	4	Industrial Design Heroes	68 Lincoln Ave.	Berkeley	CA
9	► April						
12	▼ 2010						
13	► February						
16	▼ March						
17	2010	March	12	Industrial Design Heroes	68 Lincoln Ave.	Berkeley	CA
18	2010	March	14	Fuzzy Resolution Corp.	1334 Grant Blvd.	Redding	CA
19	▼ 2011						

8.7 After organizing a table by categories, you can expand or collapse any category or subcategory.

Here's how to create categories manually. Use this technique when you want to choose the rows that the category contains.

1. **Select at least one cell in each of the rows you want to place in the category.** The rows can be next to each other, but they don't have to be. To select cells that aren't next to each other, click the first cell, and then ⌘+click each of the others.

2. **Move the mouse pointer over the reference tab for one of the selected rows, click the pop-up button, and choose Create Category from Selected Rows.** Numbers inserts a category row, which looks like a gray bar, puts the rows under it, and gives it a default name, such as Item 1.

3. **Double-click the default name, type the name you want to assign the category, and then press Return.**

Note

You can also create a new category above a row by clicking the row, moving the mouse pointer over the row header, clicking the pop-up button, and choosing Insert Category. Double-click the category's default name, type a new name, and press Return.

You can also have Numbers create categories and subcategories for you from the data in the columns you choose. Here's what to do:

1. **Click anywhere in the table in which you want to create the categories.**

2. **Click the Reorganize button on the toolbar to open the Reorganize window.**

3. **If the Categories section (the bottom section) is hidden, click the Categories disclosure triangle to display it.**

4. **In the first pop-up menu, choose the header name or column letter to use for the categories.** The Reorganize window displays a second pop-up menu.

5. **In the second pop-up menu, choose how to divide up the categories.** These are your choices:

   ○ **unique values.** This creates a category for each unique item and puts all the rows that contain the item in that column into the category. For example, say you choose column B, which contains Truck in cell B2, Car in cell B3, Bike in cell B4, and Truck in cell B5. Numbers creates three categories: Truck (with two rows), Car (with one row), and Bike (with one row).

   ○ **days.** Numbers creates a category for each day.

   ○ **weeks.** Numbers creates a category for each week.

   ○ **months.** Numbers creates a category for each month.

   ○ **quarters.** Numbers creates a category for each quarter of the year.

   ○ **years.** Numbers creates a category for each year.

6. **If you want to create a subcategory, click the + button at the right end of the first row.** Numbers adds a second row of controls.

7. **Set up the subcategory as described in Steps 4 and 5.** Figure 8.8 shows an example of setting up a subcategory.

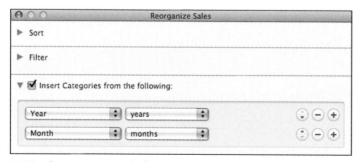

8.8 Numbers can automatically create categories and subcategories from the columns you choose.

8. **Set up further subcategories as needed.** For example, it's often useful to set up Year categories with Month subcategories and Day subcategories (within the months).

9. **Look at the categories and subcategories your choices produce in the table, and adjust them as needed.** You can move a category or subcategory up or down the list in the Reorganize window, or click the – button on its row to remove it.

When you want to view your table without the categories, choose Table ⇨ Disable All Categories or deselect the Insert Categories from the following check box in the Reorganize window. To display the categories again, choose Table ⇨ Enable All Categories or select the Insert Categories from the following check box once more.

Creating Charts from Your Table Data

In Numbers, you can insert charts in your documents using the same techniques as for Pages and for Keynote, as discussed in Chapter 1. But you may also want to take advantage of the extra charting features that Numbers offers, as discussed here.

Inserting a chart

The standard way of inserting a chart in Numbers is to select the range of adjacent cells that contains the data — or click the table handle if you want to use the whole table — and then give the command for inserting the chart. This works well when your data is neatly arranged and you want to use all of it. But Numbers enables you to use data from an entire table, from nonadjacent cells, or from two or more tables, as described next.

Note   You can insert a chart either from the Charts pop-up menu on the toolbar or from the Insert ⇨ Chart submenu. Use the Charts pop-up menu when you want to pick a chart type by its visual appearance; use the Chart submenu when you want to pick a chart by name (for example, 3D Stacked Column).

Creating a chart from nonadjacent cells

In many spreadsheets, you'll need to create a chart that uses only some of the data from a table rather than all of it. For example, you may want to omit data that detracts from the point you're trying to convey in your chart, or you may simply find that including all the data may make the chart too complex for easy reading.

You could pull out the data you want into another table, or create a copy of this table and then cut it down to only the data you want, but Numbers has a better way — just select the cells you want within the table.

Here's how to create a chart from nonadjacent cells:

1. **Click the first cell, or click and drag through the first range of cells.**

2. **Hold down ⌘ while you click each other cell you need, or while you click and drag through each other range of cells.**

Figure 8.9 shows an example of selecting two columns in a table.

	A	B	C	D	E
1	State	Ohio	Arizona	Idaho	Colorado
2	2009	49	14	19	6
3	2010	18	18	37	42
4	2011	63	73	25	18
5	2012	24	61	43	68

8.9 You can create a table from nonadjacent cells or ranges of cells.

Genius

Instead of inserting a chart of the default size and then resizing it and maybe repositioning it, you can draw a chart of your preferred size wherever you like. Select the data for the chart, then hold down the Option key while you open the Charts pop-up menu on the toolbar and click the chart type. Click and drag the resulting crosshair across the sheet canvas to place and size the chart.

You can quickly reposition a chart on its current sheet by clicking it and dragging to the new position. To move a chart from one sheet to another, cut it from the current sheet and paste it on the other sheet.

Creating a chart from two or more tables

One really handy trick is to create a chart that draws data from two or more tables. By doing this, you can avoid having to consolidate all the data into a single table.

Genius

Drawing data from different tables also lets you contrast different data sets in the same chart. For example, you could create a line chart that shows the Dow and NASDAQ plunging into the credit crisis — and then add data from a different table that shows your company's stock price soaring through the same period.

To use data from two or more tables, create the chart from the data in the first table as usual. Then click the chart to select it, hold down the ⌘ key, and then click the cells or drag through the ranges in another table to add that data.

Extending a chart with more data

Another move you may need to use with charts is extending a chart so that it displays more data than you originally used to create it. For example, you may need to add the latest sales figures to freshen up a chart. The old-fashioned approach would be to delete the chart and create it again from scratch, but Numbers is smart enough to enable you to sidestep wasting time like this.

You can extend a chart with more data in any of these ways:

- **Insert a new row or column in the table.** If the chart uses data from a single table, you can simply insert a new row or column between the existing rows or columns. Enter the data in that row or column, and Numbers updates the chart to show it.

- **Select more cells from the same table.** Click the chart to select it. Then Shift+click and drag to select cells that are adjacent to charted cells, or ⌘+click or ⌘+click and drag to select cells that are not adjacent to charted cells.

- **Drag the chart control to take in more cells.** If the chart uses only part of a table, click the chart to select it, and then drag the chart control down, to the right, or both (see figure 8.10). The chart control is the empty circle at the lower-right corner of the chart area. When you're dragging the chart control, the mouse pointer appears as a black cross.

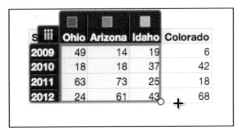

S iii	Ohio	Arizona	Idaho	Colorado
2009	49	14	19	6
2010	18	18	37	42
2011	63	73	25	18
2012	24	61	43	68

8.10 Click the chart and then drag the chart control to select more cells in the table to add their data to the chart.

Note

You can also drag the chart control up, to the left, or both to reduce the amount of data in the chart.

- **Add a data series from another table.** Select the data from which you want to create the data series, and then click and drag it to the chart.

Choosing whether to display hidden rows or columns in a chart

As you've seen in the previous chapters, you'll often need to hide rows or columns in your tables to make the tables appear the way you want them to. When you create a chart based on a table with hidden data, you need to decide whether to show that hidden data in the chart or keep it hidden.

Genius

Decide whether to include data that's hidden in the table depending on what the data shows, why it's hidden, and whether showing it will make the chart more or less persuasive. Also consider whether the audience will be able to tell (for example, from a handout) that the data was hidden in the source table; if so, using the data in the chart may be embarrassing rather than helpful.

To make a chart display the data in any hidden rows or columns in its data range, select the Show Hidden Data check box in the Chart Inspector. Deselect this check box to exclude the hidden data from the chart.

Removing values from a chart without changing the table

The easy way to remove a value from the chart is simply to delete it from the table; Numbers then removes the value from the chart as well. But sometimes you may need to retain values in the table but prevent them from appearing in the chart — for example, because the data set is anomalous or undermines your argument.

To remove a data set like this, click the chart, click the data set's category label in the table, and then press Delete. Numbers removes the data from the chart but keeps it in the table.

Linking charts

When you've finished creating a chart in Numbers, you can copy it from the sheet and paste it into a Pages document or onto a Keynote slide. When you do this, Pages or Keynote creates a link between the copy of the chart and the original in Numbers.

If you change the chart in Numbers, you can quickly update it in Pages or Keynote. Click the chart to display its control bar, and then click the Refresh button (the button with the curving arrows, shown in figure 8.11). Pages or Numbers pulls in the latest data from the saved version of the Numbers workbook.

If you need to share the Pages document or the Keynote presentation with someone who won't have the Numbers spreadsheet available, unlink the chart. Click the chart to display the control bar, click the right end of the control bar to display the Unlink button, and then click the Unlink button.

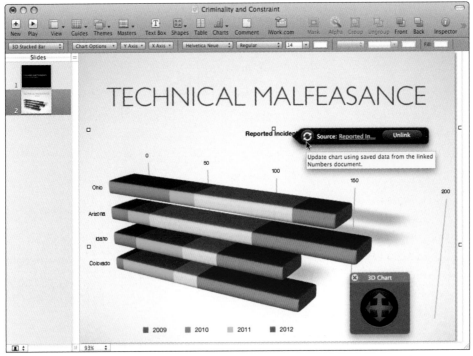

8.11 You can quickly update a chart you've placed in a Keynote presentation or a Pages document with the latest data from Numbers.

Customizing and Modifying Table Styles

The built-in table styles in Numbers' templates are great for giving a spreadsheet document an attractive and consistent look. But if you find that the built-in styles don't give the look you want, don't waste time applying manual formatting to your tables: Either create a custom table style of your own or modify an existing style to meet your needs.

Creating a custom table style

Here's how to create a custom table style:

1. **Choose a table to work in.** Either pick an existing table in the spreadsheet document, or insert a new table so that you can work freely.

2. **Apply the closest existing style.** Click in the table, and then click the style in the Styles pane.

3. **Change the style's formatting as needed.** Set the formatting for each of the table's elements — body cells, header cells, and footer cells — in turn:

   - **Format the text.** Use the controls on the Format bar or the Fonts window as usual.

   - **Format the borders.** Use the controls on the Format bar or in the Table Inspector, as discussed earlier in this chapter.

   - **Add a fill or background.** Use the Fill control on the Format bar or the Cell Background controls in the Table Inspector.

4. **In the Styles pane, highlight the style's name, click the pop-up button, and then choose Create New Style.** Type the name for the style in the Table style name dialog box (see figure 8.12), and then click OK.

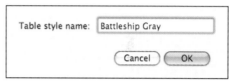

8.12 Assigning a name to a custom table style.

The new style appears in the Styles pane. To apply the style, drag it to a table or select the table and then click the style.

Note You can drag the styles in the Styles pane into a different order. For example, you can put the styles you use most frequently at the top of the pane to reduce the number of mouse miles you rack up.

Modifying an existing table style

Instead of creating a custom table style as described previously, you can modify an existing style and then save the changes with that style name. This method helps keep down the number of table styles in the spreadsheet document and also lets you apply the change instantly to all the other tables in the document that use the same style.

To save the changes you've made to a style, highlight the style's name in the Styles pane, click the pop-up button, and then choose Redefine Style from Table. Numbers changes the table style's definition and updates all the document's tables that are formatted with that style so they all take on the new look.

Setting a default table style

When you insert a new table on a sheet, Numbers automatically applies the spreadsheet document's default table style to it. You can change the default table style by highlighting the style you want in the Styles pane, clicking the pop-up button, and then choosing Set as Default Style for New Tables.

Deleting a table style

When you no longer need a table style, you can delete it by highlighting the style's name in the Styles pane, clicking the pop-up button, and then choosing Delete Style.

If any table is using the style, Numbers prompts you to choose a replacement style (see figure 8.13).

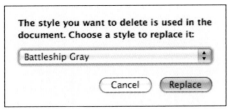

8.13 When you delete a style that a table is using, you need to choose a replacement style.

Adding Media with the Media Browser

You can add images, audio files, and movies to your spreadsheets by using the Media Browser as discussed in Chapter 1. All-singing, all-dancing spreadsheet documents tend to be less useful than multimedia presentations, so you may not need to add audio files or movie files. But images can greatly enhance a spreadsheet. This section discusses the three main ways to use them.

Placing an image on the sheet canvas

The simplest way to add an image is to place it on the sheet canvas. Click the sheet canvas, then click and drag the image from the Photos pane of the Media Browser or from a Finder window. You can then mask the image as needed, and add a frame if you want.

Adding a background image to a cell, table, or chart

You can add an image as the background to a cell, an entire table, or a chart. Click the object, open the Graphic Inspector, choose Image Fill in the Fill pop-up menu, and then choose the image.

Putting a background image behind multiple objects

If you want to place an image behind two or more objects, you need to proceed differently. Here's what to do:

1. **Place the image on the sheet canvas.** Drag it from the Photos pane of the Media Browser as usual.

2. **Resize and mask the image as needed.**

3. **Send the image behind or to the back.** Ctrl+click or right-click the image and choose Send Backward or Send to Back, as needed. Send to Back is easiest if you're just using one image. If you need to layer one image on top of another, use Send Backward.

4. **Position the tables or other objects over the object.**

5. **Adjust the opacity of the objects.** Open the Graphic Inspector, click the first object, and then drag the Opacity slider until the image shows through the object to the degree you want.

Adding Controls to Cells

Normally, someone who uses your spreadsheet interacts with it by using the keyboard or mouse to enter or change values in cells. But you can also insert four types of controls in cells to let the user perform particular actions: check boxes, sliders, steppers, or pop-up menus.

Adding a check box to a cell

Add a check box to a cell when the user needs to make a simple yes/no or true/false choice in a table. Unlike when a check box appears in a dialog box, a check box in a cell doesn't directly change any settings, but you can reference it in formulas. When a check box is selected (see figure 8.14), it returns TRUE; when it is unselected, it returns FALSE.

Here's how to add a check box to a cell:

1. **Click the cell in which you want to place the check box.**

8.14 A check box is an unambiguous way of requiring the user to make a yes/no choice.

Note Every computer user understands the meaning of a check box, so it's a good control to use when you need the user to make a choice without any possible ambiguity.

2. **Click the Checkbox button on the Format bar.** Numbers inserts a check box in the cell and displays the Cells Inspector with Checkbox selected in the Cell Format pop-up menu (see figure 8.15).

3. **Choose the check box's default state.** Select the Checked option button or the Unchecked option button, as appropriate.

Close the Cells Inspector, click the check box cell, and make sure that you can select it and deselect it.

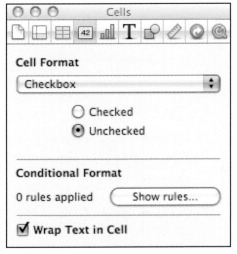

8.15 In the Cells Inspector, choose whether the check box should start off selected or unselected.

Note To remove a check box from a cell, click the cell, and then press Delete.

Adding a slider to a cell

Add a slider to a cell when you want to let the user make large changes to a value within the range you choose. For example, you can add a pricing slider that lets the user change a product's price to anywhere from $5 to $500 in five-dollar increments, as in figure 8.16.

8.16 A slider is useful when the user needs to make large changes to a value.

Here's how to add a slider to a cell:

1. **Click the cell in which you want to place the slider.**

2. **Open the Cell Formats pop-up menu on the Format bar, and then click Slider.** Numbers inserts a slider to the right of the cell and displays the Cells Inspector with Slider selected in the Cell Format pop-up menu (see figure 8.17).

3. **Set the values for the slider.** Enter the lowest value allowed in the Minimum box, the highest value allowed in the Maximum box, and the size of the steps in the Increment box.

4. **In the Position area, choose where to position the slider.** Select the Right option button to position the slider to the right of the cell you're using. Select the Bottom option button to position the slider below the cell you're using.

5. **In the Display as pop-up menu, choose how to the display the slider.** Your choices are Number, Currency, Percentage, Fraction, Scientific, or Numeral System. For example, if the user is choosing standard pricing in dollars, choose Currency. You can then choose whether to use accounting style (with the currency symbol aligned at the left of the cell) or regular style (with the currency symbol appearing just before the numeric value).

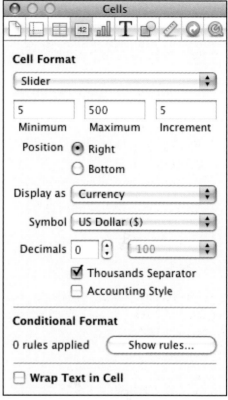

8.17 Set the range of values for the slider, choose the increment, and decide whether to position it to the right of the cell or below it.

After you finish choosing options for the slider, close the Cells Inspector. Click the slider cell, make sure the slider appears in the correct position, and then drag the slider to make sure it's using the values you selected.

Adding a stepper to a cell

Add a stepper to a cell when you want to let the user choose values in predefined increments — for example, 5, 10, 15, 20, and so on. Figure 8.18 shows a stepper in a table.

Here's how to add a stepper to a cell:

1. **Click the cell in which you want to place the stepper.**

2. **Open the Cell Formats pop-up menu on the Format bar, and then click Stepper.** Numbers inserts a stepper in the cell and displays the Cells Inspector with Stepper selected in the Cell Format pop-up menu (see figure 8.19).

8.18 Add a stepper to let the user increase or decrease the value by a set amount within a specified range of values.

3. **Set the values for the stepper.** Enter the lowest value allowed in the Minimum box, the highest value allowed in the Maximum box, and the size of the steps in the Increment box.

4. **In the Display as pop-up menu, choose how to the display the stepper.** Your choices are Number, Currency, Percentage, Fraction, Scientific, or Numeral System. For example, to create the example stepper shown, choose Number and enter 3 in the Decimals box.

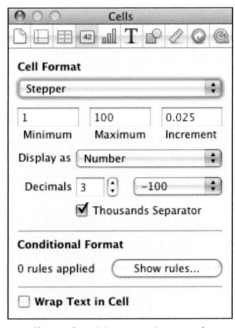

5. **Choose other formatting options as needed.** For example, set the number of decimal places, choose how to represent negative numbers, and decide whether to use the thousands separator comma.

After you finish choosing options for the stepper, close the Cells Inspector. Click the stepper cell, and then click the Up button or the Down button to make sure the stepper is working as you intended.

8.19 Choose the minimum, maximum, and increment values for the stepper.

Adding a pop-up menu to a cell

Add a pop-up menu to a cell when you want to provide the user with a limited range of values to choose from. For example, in an expense report, you may need to include a pop-up menu that lets users choose the office or department for which they work. Figure 8.20 shows an example.

A pop-up menu can help both the user and you. It helps the user by not only making clear exactly what kind of value the cell needs but also limiting the choice of values. (If none of the values seems suitable, though, the user may remain puzzled.) And it helps you by preventing the user from entering the wrong kind of value in the cell.

Here's how to add a pop-up menu:

1. **Click the cell in which you want to place the pop-up menu.**

2. **Open the Cell Formats pop-up menu on the Format bar, and then click Pop-up Menu.** Numbers inserts a pop-up menu in the cell and displays the Cells Inspector with Pop-up Menu selected in the Cell Format pop-up menu (see figure 8.21).

3. **Enter the values you want the pop-up menu to show.**

   - **Change a default value.** In the list box, double-click one of the default values (1, 2, and 3) to display an edit box around it. Type the value you want, and then press Return. Repeat the process for each of the other values.

   - **Add a value.** Click the + button, type the name for it over the default name, and press Return.

   - **Delete a value.** Click the item, and then click the – button.

After you enter all the values, close the Cells Inspector. Click the pop-up button on the cell, and make sure that the pop-up menu appears as you want it to.

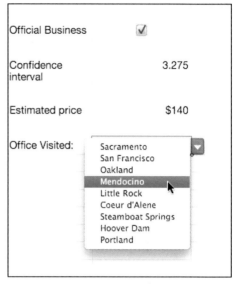

8.20 A pop-up menu restricts the user to an easy range of choices.

8.21 When you insert a pop-up menu, use the Cells Inspector to set the values it will contain.

243

How Can I Make My Spreadsheet Easy to Use and Share?

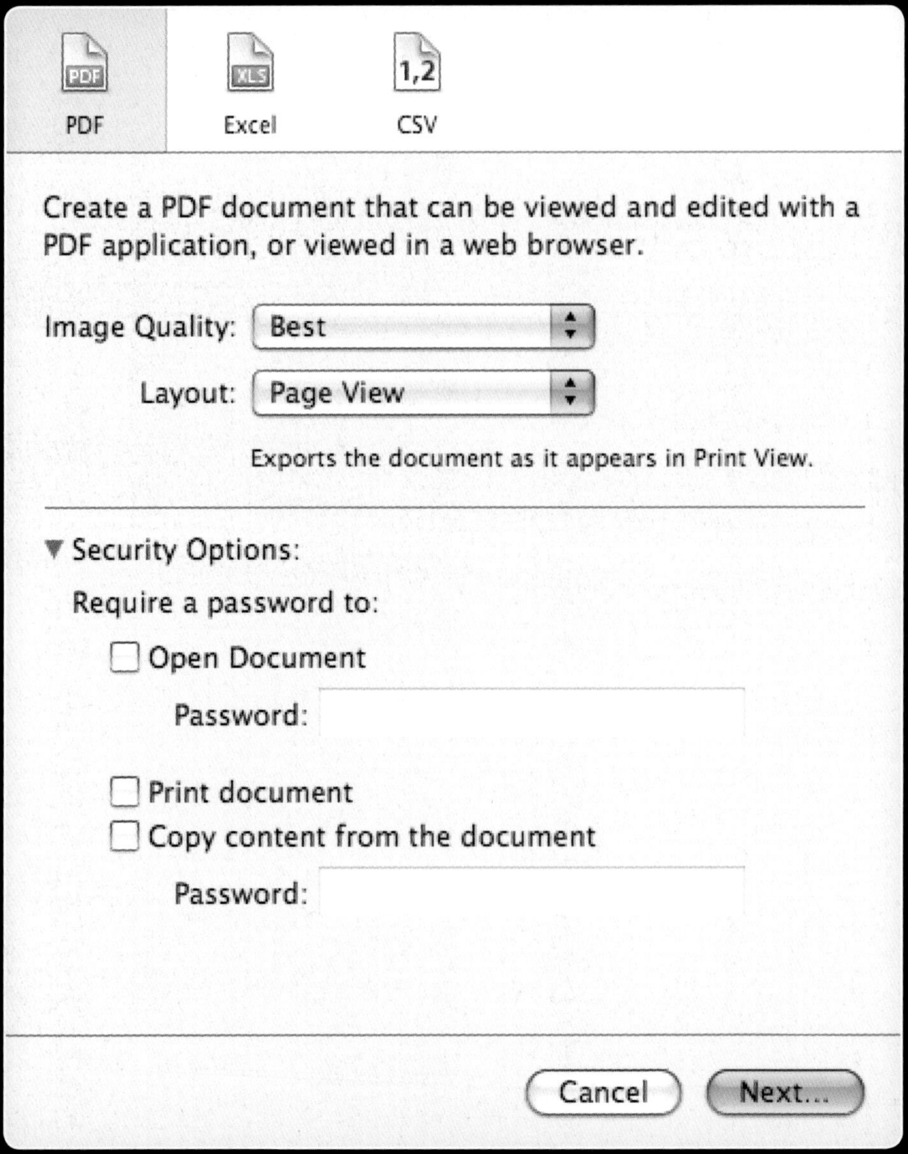

PDF	Excel	CSV

Create a PDF document that can be viewed and edited with a PDF application, or viewed in a web browser.

Image Quality: Best

Layout: Page View

Exports the document as it appears in Print View.

▼ Security Options:

Require a password to:

☐ Open Document

Password:

☐ Print document
☐ Copy content from the document

Password:

Cancel   Next...

After building a spreadsheet, you're probably ready to share it with others. Before the spreadsheet is final, you can use the comment feature to gather feedback from your colleagues or to provide explanations for complex parts. Once you've finalized the content, you can get your spreadsheet ready for distribution by adding headers, footers, and page numbers; and then share it either on paper or online using any format from a fully editable Numbers spreadsheet to a text file with no formatting.

Using Comments

Numbers' comments enable you to add information to a spreadsheet or to give your colleagues suggestions on their spreadsheets.

Here's what you can do with comments:

- **Insert a comment.** Click the cell or object to which you want to attach the comment. Click the Comment button on the toolbar or choose Insert ⇨ Comment. Numbers inserts a comment bubble and links it to an orange marker in the upper-right corner of the cell. Type or paste the text of the comment in the bubble (see figure 9.1).

	A	B	C	D	E
1	Assets	Q1	Q2	Q3	Q4
2	Current Assets	$44,000	$38,628		
3	Cash	$16,450	$10,023		
4	Accounts Receivable				
5	-Doubtful Accounts				
6	Inventory		We're looking at a severe decline in cash for several reasons: * Business is down overall * More trades and payments in kind		
7	Temporary Investments				
8	Prepaid Expenses				
9	Other Current Assets				
10	Total Current Assets	$16,450	$10,023	$0	$0

9.1 The comment bubble shown linked to a specific cell (C3). The marker in cell B2 indicates a hidden comment.

- **Resize a comment.** Drag the sizing handle at the comment's right end to the left or the right (you can't make the comment deeper without typing in returns). For a comment attached to an object, drag the handle in the lower-right corner to resize the comment both horizontally and vertically.

- **Hide a comment.** If the comment is attached to a cell, you can hide it by clicking the – button in the upper-left corner of the comment bubble. You can't hide a comment attached to an object, but you can hide all comments by choosing Hide Comments from the View pop-up menu on the toolbar or the View menu on the menu bar.

- **Display a hidden comment.** Hold the mouse pointer over the comment marker to pop the comment bubble up for a moment. Click the comment marker in the cell to display the comment bubble until you hide it.

- **Delete a comment.** Click the X button in the upper-right corner of the comment bubble.

- **Reposition a comment.** Click the comment bubble's title bar and drag the bubble to where you want it.

● **Move a comment.** The only way to move a comment is to move the cell to which it is attached; when you do this, the comment goes along for the ride. If you need to move a comment so that it refers to a different cell, copy the text of the comment, delete the comment, create a new comment, and paste in the text.

Adding Headers and Footers to a Spreadsheet

To make a spreadsheet easy to identify on printouts or in PDFs, you'll probably want to add headers and footers to the pages — text that prints at the top (header) or bottom (footer) of each page. For example, you may want to give the printouts a descriptive name, page numbers, and a date.

With Numbers, adding headers and footers takes only a moment:

1. **Choose Show Layout from the View pop-up menu on the toolbar or the View menu on the menu bar, or press ⌘+Shift+L.** Numbers switches from Normal view to Print view (unless it's already in Print view) and displays the header area (at the top of the sheet) and the footer area (at the bottom of the sheet).

2. **Click in the header area or the footer area, and then type or paste in the text you want.** Figure 9.2 shows an example of adding a header.

Header area

9.2 Once you've displayed the header area, you can simply type or paste text into it.

Note The header and footer areas in most Numbers templates are formatted with the text starting aligned at the left margin. You can press Tab to move the insertion point to the right margin and start creating right-aligned text there.

3. **Add page numbers or other document information to the header or footer as needed:**

- Position the insertion point where you want the information, open the Insert menu, and then choose Page Number, Page Count, Date & Time, or Filename as needed.

- To change the format used for a Page Number or Page Count field, select the field, Ctrl+click or right-click, and then choose the format from the shortcut menu. You can choose among 1, 2, 3; a, b, c; A, B, C; i, ii, iii; and I, II, III formats.

- To change the format used for a Filename field, select it, Ctrl+click or right-click, and choose Edit Filename Format. In the panel that appears (see figure 9.3), select the Show directory path check box if you want to see the path. Select the Always show filename extension check box

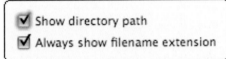

9.3 You can choose to include the directory path and filename extension with the spreadsheet's filename.

if you want to see the extension (for example, .numbers) as well as the filename. Click in the sheet to dismiss the panel.

4. **Format the header or footer text as needed.** For example, you may want to change the font, font size, or alignment. You can use either the controls on the Format bar or the Font window, as needed.

Note You can also display the header and footer areas by moving the mouse pointer over these areas in Print view. Once the outline of the header or footer area appears, click in it.

5. **When you've finished creating the header and footer, hide the header and footer areas again.** Click the Print View button at the lower-left corner of the sheet canvas or choose Hide Print View from the View pop-up menu on the toolbar or the View menu on the menu bar.

Genius You can insert the Page Number, Page Count, Filename, and Date & Time fields anywhere in a spreadsheet — they're not confined to the header and footer areas, even though this is where they're normally most useful.

Preparing a Spreadsheet for Printing and Sharing

By now, you've probably got a spreadsheet with all its contents in place and finalized. Before you print it, you will normally want to make sure you've chosen the right printer and paper size, and set the page orientation and margins the way you want them. You can then set up the spreadsheet for printing by dividing it into pages, and resizing objects if necessary.

Note

If you need to print the spreadsheet quickly in its current state, press ⌘+P to open the Print dialog box, and then click the Print button.

Setting the printer and page size

At this point, you need to tell Numbers which printer you'll use and which page size you plan to print on. Follow these steps:

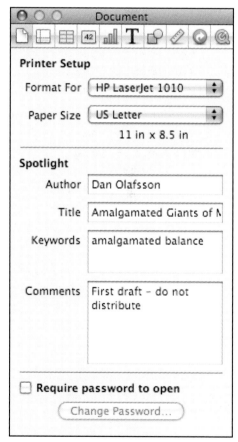

9.4 Use the Document Inspector to set the printer and paper size for your spreadsheet.

1. **Open the Document Inspector (see figure 9.4).** Click the Inspector button on the toolbar, and then click the Document Inspector button.

2. **In the Printer Setup area, choose the printer in the Format For pop-up menu.** If you know which printer you'll use to print this document, choose the printer by name. If you want to keep your options open for the time being, choose Any Printer.

3. **In the Paper Size pop-up menu, choose the paper size you will use.** For example, choose US Letter if you're using 8.5 × 11-inch paper, or choose US Legal if you're using 8.5 × 14-inch paper.

Setting the page orientation and margins

Numbers gives you an almost infinite sheet canvas on which to create your tables, so there's plenty of space for all your data. As a result, when you come to print your spreadsheets, you may find that physical paper sizes are uncomfortably small for your data.

Your first step is to set the page orientation and the margins like this:

1. **In the Sheets pane, click the sheet you want to affect.** Different sheets can have different orientations and margins.

2. **Switch to Print view.** Click the Print View button at the lower-left corner of the sheet canvas, or choose Show Print View from the View pop-up menu on the toolbar or the View menu on the menu bar.

3. **Open the Sheets Inspector (see figure 9.5).** Click the Inspector button on the toolbar, and then click the Sheets Inspector button.

4. **In the Page Layout area, click the orientation button you want.** Click the left button in the left pair to set portrait orientation; click the right button to set landscape orientation.

5. **If your sheet will print on multiple pages, use the right pair of buttons in the Page Layout area to set the order.** Click the left button to print first from the top to the bottom, and then move to the right. Click the right button to print first from left to right, and then move down.

6. **In the Page Margins area, set the distances for the left, right, top, and bottom margins.** You can also set the distance for the header and footer from the edge of the page. The top margin is the distance from the bottom of the header to the sheet's content; the bottom margin is the distance from the top of the footer to the sheet's content.

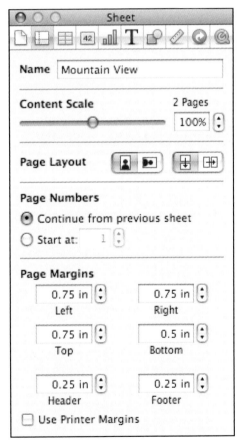

9.5 In the Sheets Inspector, you can set page orientation, margins, and page numbering.

Genius

You can also set the orientation quickly by clicking the Portrait button or the Landscape button to the right of the Print view button. Numbers shows these buttons only when you're using Print view. The button for the current orientation appears in blue.

Note

If you want to print as close to the edge of the paper as possible so that you can see your spreadsheet as large as it will go, select the Use Printer Margins check box. Numbers applies those margins and makes the Left, Right, Top, and Bottom boxes unavailable. You can still set the Header and Footer margins.

Dividing a sheet into pages for printing

To get a good-looking printout, you'll often need to divide a sheet into separate pages like this:

1. **Make sure you've set the printer and paper size, as described earlier in this chapter.**

2. **Switch to Print view.** The easiest way is to click the Print View button near the lower-left corner of the sheet canvas.

3. **Adjust the zoom level so that you can see the number of pages you want.** Click the Zoom pop-up menu at the lower-left corner of the sheet canvas and choose a zoom percentage. Alternatively, zoom out by pressing ⌘+< or choosing View ⇨ Zoom ⇨ Zoom Out, or zoom in by pressing ⌘+> or choosing View ⇨ Zoom ⇨ Zoom In.

4. **Drag the Content Scale slider at the bottom of the Numbers window to zoom all of the sheet's contents at once to a suitable size.** In figure 9.6, the sheet named Balance Sheet contains four tables that will print on four pages, but at the setting shown, parts of the second, third, and fourth tables will be broken across pages. By dragging the Content Scale slider to the right, you can increase the size of the tables so that each occupies its own page. (You may need to fine-tune the placement, as described next.)

5. **After setting the content scale, click and drag any objects you still need to move.** If an object is in the wrong place, select it, and then click and drag it to the right place. For example, click the table handle to select an entire table, move the mouse pointer over a table border so that it appears as a four-headed arrow, and then click and drag the table to a page.

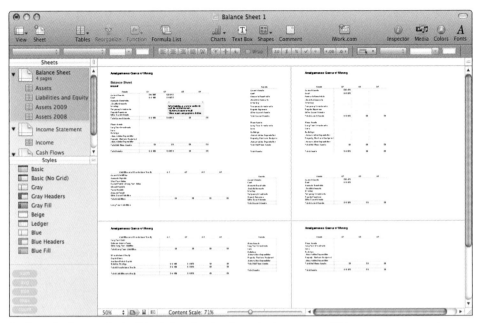

9.6 Drag the Content Scale slider to break a sheet's content as you want it to appear on separate pages.

Printing All or Part of a Spreadsheet

When you've arranged the spreadsheet into pages and divided them up for printing, you can quickly print. Here's what you need to do:

1. **If you want to print only one sheet from the spreadsheet, click it in the Sheets pane.** If you want to print all the sheets in the spreadsheet, select whichever sheet you'll find most helpful to preview.

Note If you want to include some or all comments in the printout, make sure the comments you want to print are displayed. Numbers then prints them when you print the sheet.

2. **Open the Print dialog box (see figure 9.7).** Press ⌘+P or choose File ➪ Print. If the Print dialog box opens at its small size, click the disclosure triangle to the right of the Printer pop-up menu to display the rest of the dialog box.

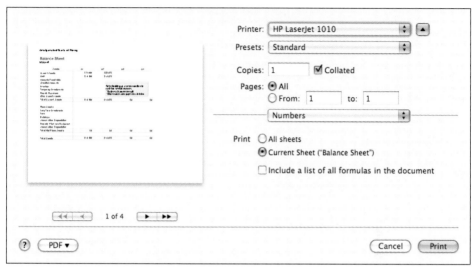

9.7 In the Print dialog box, choose whether to print all sheets or just the current sheet. You can also choose to print a list of the formulas in the spreadsheet.

3. **Use the preview on the left to make sure the sheet appears the way you want it.** Click the buttons below the preview to move among its various pages.

4. **In the Print area, select the All sheets option button if you want to print all the sheets in the spreadsheet.** Select the Current Sheet option button if you want to print just the sheet you selected in Step 1.

5. **If you want to print out a list of the document's formulas, select the Include a list of all formulas in the document check box.**

6. **Click the Print button.**

Sharing a Spreadsheet

When you've finished and polished a spreadsheet, you'll probably want to share it with others, either locally in your office or via the Internet. Numbers lets you easily share a document by placing it on iWork.com; by sending it to iWeb for download as either a Numbers file, an Excel workbook file, or a PDF file; or by exporting it as a PDF file, an Excel workbook, or a Comma-Separated Values (CSV) file.

Table 9.1 summarizes your different options for sharing spreadsheets and suggests when to use each.

Table 9.1 Ways of Sharing Your Spreadsheet

Document Type	Share Your Document in This Way When
Share on iWork.com	You want to make the spreadsheet available for other iWork.com users to comment on or download and use.
Send via Mail	You want to send the spreadsheet via email, either in Numbers format or as an Excel workbook or a PDF file.
Send to iWeb as a Numbers document	You want visitors to your iWeb Web site to be able to download the spreadsheet and edit it in Numbers.
Send to iWeb as a PDF file	You want visitors to your iWeb Web site to be able to view the spreadsheet exactly as you created it, but not be able to edit it.
Create a PDF file	You need to share the spreadsheet electronically, keeping the look and layout you've given it in Numbers.
Create an Excel file	You need to share the spreadsheet with people who use Microsoft Excel.
Create a CSV file	You want to create a version of the spreadsheet's data that's compatible with any spreadsheet application or text editor.

Before you share a spreadsheet in any of these ways, open it if it's not already open. If it is open, save any unsaved changes by pressing ⌘+S or choosing File ⇨ Save. You may also want to protect the spreadsheet with a password as discussed in Chapter 1 to prevent anyone unauthorized from opening it easily.

Sharing a document on iWork.com

When you're collaborating with other users of Apple's iWork.com online service, iWork.com is a great way of sharing Numbers spreadsheets. See Chapter 1 for the details on how this process works. You can share a Numbers spreadsheet in any or all of four formats: Numbers '09 format, Numbers '08 format (for anyone who hasn't upgraded to iWork '09), Excel workbook format, or PDF.

Sending a document via email

You can attach a Numbers spreadsheet to an email message just as you can any other file, but you can also quickly share it by choosing Share ⇨ Send via Mail and then selecting the format from the Send via Mail submenu: Numbers, Excel, or PDF.

Caution

Before you use the Share ⇨ Send via Mail ⇨ Excel command, export your spreadsheet to an Excel workbook manually and check that it comes out correctly. If you don't check, you could export and send a workbook that contains either minor or major errors and remain completely unaware of them.

If Numbers crashes when you use the Share ⇨ Send via Mail ⇨ Excel command, look to see if you still have iWork '08 on your Mac; if so, you may need to remove iWork '08 to make this command work. If you don't want to get rid of iWork '08, export the spreadsheet to Excel format (as described next) and then attach it to an email message manually. Better still, check the Excel workbook before attaching it to the message.

Sending a spreadsheet to iWeb

Numbers enables you to send a spreadsheet to a blog or podcast on your iWeb site so that visitors to the site can download the spreadsheet.

1. **Choose File ⇨ Send to iWeb, and then choose the format from the submenu:**

   - **Numbers.** Choose this format when the people who download the spreadsheet will need to edit it using Numbers. People will have full access to your data, including the formulas you've used.

   - **PDF.** Choose this format if you want people to be able to view the data but not edit it or see the formulas.

2. **Numbers opens iWeb (or activates it if it's already open).** iWeb displays the Which blog do you want to send the files to? dialog box (see figure 9.8).

3. **In the Blogs pop-up menu, choose the blog or podcast, and then click OK.** iWeb creates a page for the document. You can then fill in the placeholders as usual.

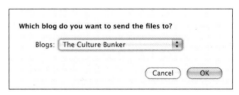

9.8 Choose which iWeb blog or podcast you want to send the Numbers document or PDF to.

Note

Numbers doesn't let you save a single sheet — or a range on a single sheet — as a Web page the way that Excel does. The only workaround is to create another Numbers spreadsheet that contains only the data you want to share.

Exporting a spreadsheet to Microsoft Excel format

When you need to share a Numbers spreadsheet with someone who uses Microsoft Excel (on either the Mac or the PC), you can save the spreadsheet as an Excel workbook with just a few clicks. But as mentioned in Chapter 6, the conversion between Numbers and Excel isn't perfect; just as some items may move and some data and formatting may disappear when you open an Excel workbook in Pages, so may a Numbers spreadsheet change when you export it to an Excel workbook.

Caution When you export a Numbers spreadsheet to Excel, always open it in Excel and check it to make sure that it has translated successfully before you send it to anyone else. Spreadsheets can become extremely complex, and moving them to a different format often introduces errors. While it's no great loss if a text box moves in a Pages document when you export it, if a formula goes wrong when converting from Numbers to Excel, it may take ages to track down the problem — and if you don't catch the error, it may prove expensive for you and your company.

Dire warnings aside, exporting a spreadsheet from Numbers to Excel takes only moments:

1. **Choose Share ⇨ Export to open the Export dialog box.**

2. **Click the Excel button to display the Excel pane (see figure 9.9).**

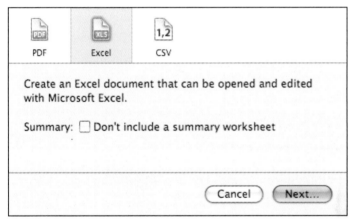

9.9 The Excel pane of the Export dialog box lets you choose whether to include a summary worksheet for spreadsheets that include multiple sheets.

3. **Select the Don't include a summary worksheet check box if you want to prevent Numbers from including a summary worksheet.** This worksheet is often helpful for spreadsheets that include multiple sheets, but you'll probably want to see if it's good for your data.

4. **Click the Next button to display the Save As dialog box.**

5. **Choose the name and folder for the exported document, and then click the Export button.** Numbers displays a readout of its progress.

If Numbers notes any problems with the export, it displays the Document Warnings dialog box to show you details. Figure 9.10 shows an example of the Document Warnings dialog box. You can click a warning to see where it occurs, and then click the Clear button to clear it; or click the Clear All button to clear all the warnings.

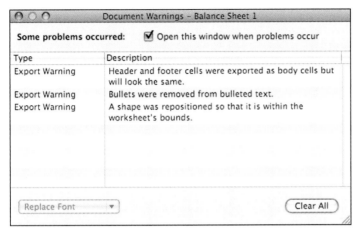

9.10 Numbers alerts you to any problems it notices when exporting a spreadsheet to Excel workbook format.

Note

You can also create an Excel workbook from a Numbers spreadsheet by choosing File ➪ Save As, selecting the Save copy as check box, choosing Excel Document in the pop-up menu, and then clicking Save. You don't, however, get to choose the filename and folder for the exported workbook: Numbers simply gives it the same name and places it in the same folder as the original.

Creating a PDF file of a spreadsheet

When you want to show other people all the content of your spreadsheet but not the formulas that make it work, create a PDF file to give to them. You can choose between displaying the spreadsheet in Page view, so that other people see the pages as you've laid them out, or in Sheet view, in which each sheet appears as a separate page.

Here's how to create a PDF file from a Numbers spreadsheet:

1. **Choose Share ➪ Export to open the Export dialog box.**

2. **Click the PDF button to display the PDF pane.** The screenshot on the opening page of this chapter shows the PDF pane with the Security Options area expanded.

3. **In the Image Quality pop-up menu, choose the image quality you want: Good, Better, or Best.**

- Best is normally what you'll want, because it produces a full-quality PDF. Numbers keeps the spreadsheet's images at their full resolution.

- If you find that Best produces files that are too large for the way you're using to distribute them, experiment with the Better setting or the Good setting to produce a smaller file. Better reduces the image quality to 150 dots per inch (dpi); Good uses 72 dpi.

4. **In the Layout pop-up menu, choose how to export the spreadsheet.**

- **Page View.** Choose Page View to create a separate page from each of the pages you've set up in Print view (as discussed earlier in this chapter). This is usually the best choice when you need to show the spreadsheet's content.

- **Sheet View.** Choose Sheet View to simply create a page from each sheet. Sheet View is useful when you want people to look at the spreadsheet itself rather than focus on its content.

5. **If you want to secure the PDF file, click the Security Options disclosure triangle to reveal the security options.** You can then require a password to open the document or a (different) password to print it or copy information from it.

6. **Click Next.** Numbers displays a Save dialog box that lets you choose the name to give the PDF and the folder in which to save it.

7. **Click Export.** Numbers displays a progress readout as it exports the PDF.

After Numbers finishes the export, open a Finder window to the folder in which you saved the PDF. Click the file, check its file size to make sure it's small enough for how you want to distribute it, and then double-click the file to open it in Preview (or your default PDF viewer). Make sure that the spreadsheet pages appear the way you want them, and that any security options you chose are working effectively, before you distribute the file.

Exporting a spreadsheet to a Comma-Separated Values file

When you need to export a spreadsheet in a format that's compatible with as many other applications and operating systems as possible, export it as a Comma-Separated Values (CSV) file. The CSV file retains only the text values from the spreadsheet — there's no formatting, no charts, no graphics, and no other objects.

In a CSV file, the value of each cell is separated from the value of the next cell with a comma. If a cell contains one or more real commas, the cell's contents are enclosed in double quotation marks to show that the comma isn't one of the separators.

Here's how to create a CSV file from a Numbers spreadsheet:

1. **Choose Share ➪ Export to open the Export dialog box.**

2. **Click the CSV button to display the CSV pane (see figure 9.11).**

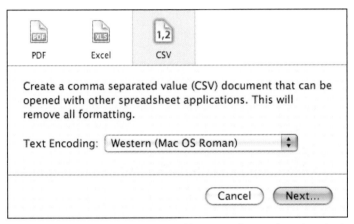

9.11 Exporting a spreadsheet to a CSV file strips it down to bare data that's compatible with almost every spreadsheet application.

3. **In the Text Encoding pop-up menu, choose the text encoding to use for the text file.** Normally, the best choice is Unicode (UTF-8), because almost all computers and operating systems can read it. If you will use the text file only on the Mac, you can choose Western (Mac OS Roman) instead. For Windows, you have the option of using Western (Windows Latin 1).

4. **Click the Next button.** Numbers displays a Save dialog box.

5. **Type the filename, and choose the folder in which to store the file.** Numbers automatically assigns the file extension .csv (which identifies a CSV file).

6. **Click Export.**

Creating Tab-Separated Values Files

CSV files are widely used for transferring spreadsheets, but so is another text-based format: Tab-Separated Values, or TSV. Instead of using a comma to separate the contents of each cell, a TSV file uses a tab. The result is easier for humans to read because of the white space between items; computers don't usually care about the difference, which is perhaps why Numbers doesn't let you create tab-separated files directly.

If you do need a TSV file, you'll need to create one manually. For straightforward data, the easiest workaround is to copy the table from Numbers, paste it into Pages, and then export the Pages document as a text file (see Chapter 5).

If you have Excel, export the spreadsheet to Excel, and then save it as a TSV file from there by opening the Save As dialog box, and then choosing Text (tab delimited) in the Specialty Formats area.

How Can I Create Presentations Quickly in Keynote?

Keynote is a fantastic tool for creating colorful and convincing presentations and delivering them to your audience in person or via the Internet. To create presentations swiftly and efficiently, you need to understand Keynote's components and what they do, set essential preferences that make Keynote look and behave the way you want, and customize the Keynote window so your essential tools are at hand. Then you can start creating your presentation — either from scratch, from an imported PowerPoint presentation, or as an outline from a document — and customizing the slides it contains.

Knowing What You Are Working With

When you open Keynote, the application automatically displays the Theme Chooser (see figure 10.1) so that you can pick a theme for a new presentation. The Theme Chooser works in a similar way to the Template Chooser in Pages and Numbers but also lets you choose the size for the slides you're creating.

10.1 In the Theme Chooser, click the theme you want to use for your presentation and choose the size for its slides.

Note If you want to open a presentation you've worked on recently rather than create a new one, click on the Open Recent pop-up menu in the Theme Chooser and choose the presentation you want to use. If you need to open a presentation you haven't worked on recently, click the Open an Existing File button, select the presentation in the Open dialog box that appears, and then click Open.

Choosing your theme and slide size

In the Theme Chooser, move the mouse pointer over a theme's thumbnail, and then move it around over the thumbnail to scroll through the various master slides in the theme. If you want to enlarge the thumbnails, drag the Size slider at the bottom of the Theme Chooser to the right. Drag the slider all the way to the left if you want to see more of the available themes at once, which can be useful when you're comparing them.

In the Slide Size pop-up menu, choose the size of the slides you want to create for your presentation. Table 10.1 explains the choices and when to use them.

Table 10.1 Keynote's Slide Sizes and When to Use Them

Resolution	Suitable for Viewing on	When to Use It
800 × 600	Many projectors, especially older ones	When using an older external projector
1024 × 768	Many projectors, especially newer ones, and standard monitors	When using a newer external projector
1280 × 720	MacBook, MacBook Air, 15-inch MacBook Pro	For presentations using your Mac's screen
1680 × 1050	17-inch MacBook Pro model, 20-inch iMac	For presentations using your Mac's screen or an external LCD (liquid crystal display) monitor
1920 × 1080	17-inch MacBook Pro, 24-inch iMac, 23-inch Apple Cinema Display, 24-inch Apple LED Cinema Display	For presentations using your Mac's screen or an external LCD monitor

If you know which monitor you will use to give the presentation, you're in a great position because you can choose the best slide size right from the start. This means that you can position the elements optimally on the slides and be sure that the slides will be pin sharp when you show them.

Often, though, you'll create a presentation without knowing which monitor it'll appear on — or maybe even who will deliver the presentation. In this case, expect the worst and plan accordingly. Rather than create a high-resolution presentation and risk having it mangled if it is shown on wretched, old hardware, go with a conservative size such as 1024 × 768 or even 800 × 600. That way, you can be sure your presentation will look okay, even if it doesn't look as great as it might have on newer hardware.

When the monitor you're using has a higher resolution than your slides, you have two choices:

- **Scale the slides up.** Keynote can adjust the slides to match the size of the monitor. Photo quality suffers a bit, and video quality suffers noticeably — but your slides appear at the full size available.

- **Stay at the original size.** Keynote simply displays a black border around the unused part of the monitor. This doesn't look great, but your slides appear at their full sharpness, and there are no problems with photo quality and video quality.

When you've chosen the theme and the slide size, click Choose. (You can also double-click the theme.) Keynote creates a new presentation based on the theme and displays the first slide.

Note

When you run a presentation on a monitor that's smaller than the slides' size, Keynote automatically scales the slides down to fit the monitor. Video quality may suffer, but this is much better than having parts of your slides — including essential contents — hanging invisibly off the top, bottom, or sides of the screen.

Navigating the Keynote window

With a new presentation open, you'll see a window that looks something like the one shown in figure 10.2.

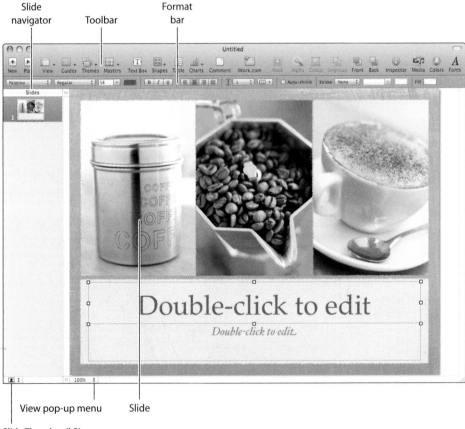

Slide navigator Toolbar Format bar

View pop-up menu Slide

Slide Thumbnail Size pop-up menu

10.2 The Keynote window with a new presentation open

As usual, the toolbar gives you quick access to the most widely used commands and features — for example, creating a new slide, setting a presentation playing, or displaying tools such as the Inspector window, Media Browser, and Fonts window. You can customize the toolbar as discussed in Chapter 1, adding the items you need and removing those you don't.

The Format bar puts the most widely used formatting for the current object immediately at hand. The contents of the Format bar change to suit the object you've selected. For example, when you select text, the Format bar gives you buttons and pop-up menus for changing the font, font face, size, and color (among other choices); when you select an image, the Format bar offers tools for changing its frame, opacity, masking, and colors.

The Slide area shows the current slide. You can zoom in and out using the View pop-up menu near the lower-left corner of the window.

The slide navigator displays thumbnails of all the slides in the presentation, letting you quickly jump from slide to slide by clicking the one you want to see. You can change the size of the thumbnails by clicking on the Slide Thumbnail Size pop-up menu and choosing Small, Medium, or Large.

At this point, you can create your first slide by clicking in a placeholder and typing text, or dragging images, sounds, and movies from the Media Browser to placeholders on the slide. But first it's usually helpful to set Keynote's preferences, as discussed next.

Setting Keynote-Specific Preferences

To make Keynote look and behave the way you want it to, spend a few minutes setting preferences. In addition to the common preferences discussed in Chapter 1, Keynote adds several General preferences and Rulers preferences; Slideshow preferences for controlling what the audience sees; Presenter Display preferences for controlling what the presenter sees; and Remote preferences for setting up an iPhone or iPod touch as a remote control for a presentation.

This section discusses Keynote's extra General preferences and Rulers preferences. Chapter 12, which explains how to give a presentation with Keynote, covers the Slideshow preferences, Presenter Display preferences, and Remote preferences.

Choose Keynote ⇨ Preferences to open the Preferences window, and then work through the following sections, choosing settings that suit you.

Setting General preferences

Start by clicking the General tab to display the General pane of the Preferences window (see figure 10.3). These are the General preferences that Keynote has but Pages and Numbers do not:

- **Reduce placed images to fit on slides.** Select this check box if you want Keynote to automatically reduce the size of any image you place that is larger than the slide on which you place it. This setting is usually helpful, especially as most digital cameras now take photos that are far higher in resolution than Keynote's slides. After Keynote automatically resizes the image, you can resize it further as needed.

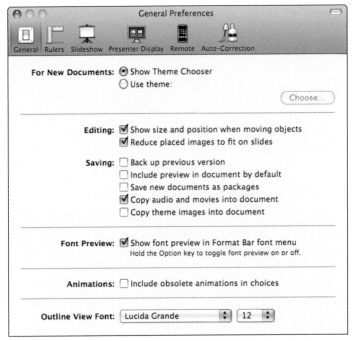

10.3 In the General pane of the Preferences window, choose whether to let Keynote resize images automatically; whether to copy audio, movie, and theme files into the presentation; and which font to use in Outline view.

- **Copy audio and movies into document.** Select this check box to make Keynote copy audio files and movie files into presentations when you save them. This is usually a good idea, because it means that the presentation file contains everything needed for the presentation, but it can increase the file size by a huge amount. If you'll deliver the presentation using the same Mac on which you create the presentation, you can deselect this check box and have Keynote use the audio and movie files from their current folders.

Note

The first time you save a presentation, you can override your preferences for copying audio, movies, and theme images. Open the Save As dialog box (choose File ⇨ Save or press ⌘+S), expand the Advanced Options area, and then select or deselect the Copy audio and movies into document check box and deselect the Copy theme images into document check box.

Copy theme images into document. Select this check box if you want Keynote to add the theme images to the presentation file. You need to do this only when you're distributing a presentation file for use without Keynote. If you're running the presentation from Keynote, the theme files should be installed on the computer.

Include obsolete animations in choices. Select this check box if you want Keynote to include older animations that Apple has phased out in Keynote '09. These include popular animations such as Confetti, Fall, and Swoosh.

Outline View Font. In the first pop-up menu, choose the font to use for outline view. In the second pop-up menu, choose the point size. Typically, you'll want a font and font size that enable you to work quickly and accurately.

Setting Rulers preferences

Keynote's Rulers preferences have many of the same settings as Pages and Numbers (see Chapter 1) but they also have two extra settings:

Master Gridlines. Click the color button and select the color you want to use for the master gridlines. To control where the gridlines appear, you can click in the Show Horizontal Gridlines Every check box or the Show Vertical Gridlines Every check box (or both) and set the percentage in the appropriate box. The default setting is 10%, but you may want to try decreasing it to 5% for fine placement. Keynote pops the gridlines up on-screen as you move an object rather than displaying them all the time, which can be distracting.

Object Spacing & Sizing. Click the color button and select the color you want to use for Keynote's spacing and sizing lines, and then choose which lines to display by selecting the Show relative spacing check box, the Show relative sizes check box, or both. Keynote displays these lines momentarily as you reposition or resize objects so that you can see how they're related to other objects.

Note You can quickly override the Alignment Guides and Object Spacing & Sizing preferences you've chosen in the Rulers pane of the Preferences window by using the options on the Guides pop-up menu on the toolbar.

Applying Themes and Master Slides

To make your presentations and slides look great, you need to understand how themes and master slides work, and how you use them. Here's what they are:

- **Master slide.** A *master slide* is a layout of text and objects for a slide. For example, the master slide for a title slide may include a text box formatted with a large font size for the title of the presentation and a second text box formatted with a smaller size for the subtitle. A master slide for an information slide may contain a title across the top, a box containing bulleted text on the left, and a placeholder for a picture on the right.

- **Theme.** A theme is a set of master slides that are related in design. For example, most themes use similar backgrounds, colors, and fonts to give a "family feel" to the master slides.

To control how a presentation looks overall, you apply a theme to it. Normally, you apply the theme by using the Theme Chooser window when you start creating the presentation, as described earlier in this chapter.

After you apply the theme, you can access the slide masters in the theme from the Masters pop-up menu on the toolbar. To apply a master to a slide, click the slide in the slide navigator, open the Masters pop-up menu, and click the master you want.

For many presentations, you will need only the masters in the theme you've chosen. For others, though, you may want to use two or more different themes. You can apply a different theme to a slide by selecting the slide, opening the Themes pop-up menu (shown on the opening page of this chapter), and choosing the theme you want.

Genius

If you want to see the full range of masters within the themes, click Theme Chooser at the top of the Themes pop-up menu to open the Theme Chooser window. You can then scroll the mouse pointer over one of the thumbnails to display the various masters. To apply a theme, click it, and then click Choose.

When you've applied two or more themes to the same presentation, the Masters pop-up menu shows both the themes and the masters (see figure 10.4).

You can reapply a slide's current master by choosing Format ▷ Reapply Master to Slide. In the Navigator, you can also Ctrl+click or right-click the slide in the Navigator and choose Reapply Master to Slide.

10.4 Choose the master slide from the Masters pop-up menu. When you've applied multiple themes to a presentation, select the theme that contains the master you want.

Setting Up the Keynote Window for Working Easily

To work quickly and smoothly in Keynote, you'll most likely want to customize the window.

First, customize the Keynote toolbar using the techniques discussed in Chapter 1. After that, adjust the zoom, find out which view suits you best for which task, and pick the best mix of screen elements, as described here.

Zooming in and out

To zoom in and out on the slide, click on the Size pop-up menu and choose the size: 25%, 50%, 75%, 100%, 125%, 150%, 200%, 300%, 400%, or Fit in Window. If you want to see the whole slide as large as possible, use Fit in Window.

You can also zoom in and out by using the View ⇨ Zoom submenu, which contains Zoom In, Zoom Out, Actual Size, and Fit in Window commands. What's more useful is zooming with keyboard shortcuts:

- **Zoom in.** Press ⌘+>.
- **Zoom out.** Press ⌘+<.

Each press of the shortcut takes you to the next or previous zoom increment on the Size pop-up menu. For example, pressing ⌘+> from 100% zooms in to 125%, while pressing ⌘+< zooms out to 75%.

Choosing the right view for each task

Keynote offers you four different ways to view your slides. You can change the view by clicking the View pop-up menu button on the toolbar or the View menu on the menu bar and then choosing Navigator, Outline, Slide Only, or Light Table.

- **Navigator view.** Navigator view displays the Slides pane on the left of the window with a thumbnail of each slide. You can change the size of the thumbnails by opening the Size pop-up menu in the lower-left corner of the Keynote window and choosing Small, Medium, or Large. This is the view in which Keynote first displays a presentation and it's the view you'll probably use most of the time when building a presentation.

- **Outline view.** Outline view replaces the Slides pane on the left of the window with an outline of the text on each slide. Figure 10.5 shows an example of outline view. Use outline view when you're focusing on the text of the presentation. Outline view displays the slide text using the font and size you chose in the Outline View Font pop-up menus in the General pane of the Preferences window.

- **Slide only view.** Slide only view hides the Slides pane or Outline pane to give you more room to view and edit a single slide. Use slide only view to work on a single slide rather than on the presentation as a whole or on the outline.

- **Light table view.** Light table view, as shown in figure 10.6, shows thumbnail versions of the slides spread out across a light table (the backlit, translucent glass table that photographers use for viewing photographic slides or negatives). Use light table view for rearranging the slides in your presentation or getting an overview of all the slides at once. In light table view, you can change the size of the thumbnails by clicking in the Size pop-up menu in the lower-left corner of the window and choosing Small, Medium, or Large.

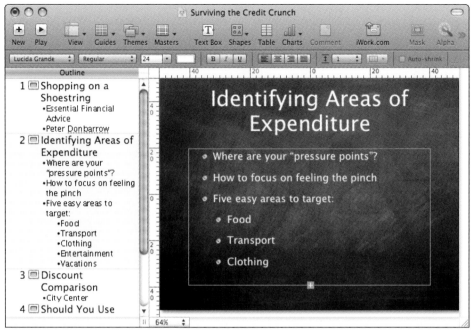

10.5 Outline view lets you quickly develop the text content of your presentation and rearrange slides into the right order. You can double-click a slide's icon to collapse it to just the title or to expand it again.

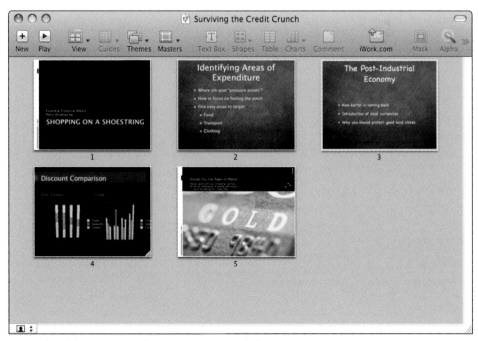

10.6 Light table view is great for reorganizing your slides.

Choosing which screen elements to display

Keynote lets you display or hide various parts of its window to suit the presentation you're working on and the tasks you're performing. Figure 10.7 shows the parts of the Keynote window, which are discussed next, that you can display or hide as needed.

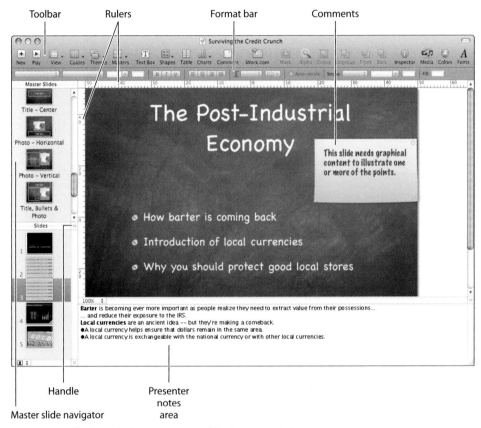

10.7 You can display or hide various parts of the Keynote window.

Toolbar

Most people find the toolbar handy for instant access to essential features, but you can also hide the toolbar if you need more screen space. To display or hide the toolbar, use one of these options:

- **Mouse.** Click the jellybean-shaped button at the right end of the title bar.
- **Keyboard.** Press ⌘+Option+T.
- **Menu bar.** Choose View ⇨ Hide Toolbar or View ⇨ Show Toolbar.

Format bar

Appearing under the toolbar (when the toolbar is displayed), the Format bar is the quickest and easiest way to apply essential formatting. But if you want, you can toggle the Format bar on and off using one of these options:

- **Mouse.** Click on the View pop-up menu on the toolbar and choose Hide Format Bar or Show Format Bar.

- **Keyboard.** Press ⌘+Shift+R.

- **Menu bar.** Choose View ⇨ Hide Format Bar or View ⇨ Show Format Bar.

Rulers

When you need to position objects exactly, display the horizontal ruler and vertical rule using one of these options:

- **Mouse.** Click on the View pop-up menu on the toolbar and choose Show Rulers or Hide Rulers.

- **Keyboard.** Press ⌘+R.

- **Menu bar.** Choose View ⇨ Show Rulers or View ⇨ Hide Rulers.

Comments

To view the comments added to a presentation, click on the View pop-up menu on the toolbar or the View menu on the menu bar and choose Show Comments. To hide the comments again, click on the menu and choose Hide Comments.

Presenter Notes area

To add notes for the presenter to a slide, open the Presenter Notes area at the bottom of the Keynote window by clicking on the View pop-up menu on the toolbar or the View menu on the menu bar and choosing Show Presenter Notes. You can then type or paste in text as needed and format it by using the Text Inspector or the Fonts window.

When you no longer need the Presenter Notes area, click on one of these View menus again and choose Hide Presenter Notes.

Master slide navigator

When you need to create a new master slide or duplicate an existing one (see Chapter 11), you can display the master slide navigator by choosing Show Master Slides from the View pop-up menu on the toolbar or the View menu on the menu bar. If you have the slide navigator open, you can also open the master slide navigator by dragging down the handle at the right end of the title bar of the slide navigator (the bar with Slides written on it).

When you've finished using the master slide navigator, either drag the handle up so that the slide navigator covers the master slide navigator, or choose Hide Master Slides from the View pop-up menu on the toolbar or the View menu on the menu bar.

Opening Microsoft PowerPoint Presentations

However much you love Keynote, chances are that you'll need to work with people who use PowerPoint, the Microsoft Office presentation application.

Keynote is largely compatible with PowerPoint — Keynote can open PowerPoint files and can export presentations as PowerPoint files — but as with Pages and Numbers, the devil is in the details. The more complex the presentation, the more likely you are to run into problems when importing it into Keynote or exporting it in PowerPoint format.

To open a PowerPoint presentation in Keynote, choose File ⇨ Open, click the PowerPoint file in the Open dialog box, and click Open. Keynote automatically converts the contents of the file to Keynote's equivalents and then displays the results.

Note   Keynote can open presentations in the PowerPoint 2007 (Windows) and PowerPoint 2008 (Mac) documents formats, which use the .pptx file extension, as well as presentations in the formats used by PowerPoint 2003 (Windows) and PowerPoint 2004 (Mac) and earlier versions, which use the .ppt file extension.

Keynote manages to import most PowerPoint slides pretty well, but the process isn't perfect. This isn't surprising given the differences between the applications, but it means you'll often need to straighten things out after importing a presentation.

If there are problems, Keynote displays the Some warnings occurred. Would you like to review them now? dialog box (see figure 10.8). You can dismiss this dialog box by clicking Don't Review and review the warnings later by choosing View ⇨ Show Document Warnings, but usually it's best to look at them immediately and decide which (if any) you need to fix — either in Keynote or by going back to PowerPoint (assuming you have PowerPoint).

Some warnings occurred. Would you like to review them now?

You can review warnings at any time by choosing View > Show Document Warnings.

Don't Review   Review

10.8 If Keynote prompts you to review warnings, have a look right away to find out which PowerPoint features it wasn't able to handle.

When you click the Review button (or choose View ➪ Show Document Warnings later), Keynote displays the Document Warnings dialog box, (see figure 10.9).

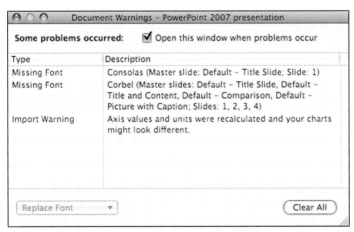

Document Warnings – PowerPoint 2007 presentation

Some problems occurred: ☑ Open this window when problems occur

Type	Description
Missing Font	Consolas (Master slide: Default – Title Slide; Slide: 1)
Missing Font	Corbel (Master slides: Default – Title Slide, Default – Title and Content, Default – Comparison, Default – Picture with Caption; Slides: 1, 2, 3, 4)
Import Warning	Axis values and units were recalculated and your charts might look different.

Replace Font ▾   Clear All

10.9 Open the Document Warnings dialog box to see a list of the problems Keynote has found when converting a PowerPoint presentation.

Click a problem in the list to see the slide on which the problem occurs. For a Missing Font problem, click the problem in the list, and then choose the replacement font from the Replace Font pop-up menu. Keynote replaces the font wherever it occurs throughout the presentation.

When you've dealt with a problem, or decided not to deal with it, click the Clear button to remove the problem from the list. To wipe the slate clean of warnings, click the Clear All button that appears when you have not selected a problem in the list.

The three areas where you're likely to have problems when you bring a PowerPoint presentation into Keynote include:

- **Charts.** Charts are staples of business presentations, and Keynote does its best to preserve PowerPoint's charts. On the plus side, Keynote does give you a version of the chart that you can edit with the Chart Data Editor. On the minus side, Keynote sometimes can't get PowerPoint's data series and axes entirely straight. So whenever you use Keynote to open a PowerPoint presentation that contains one or more charts, check immediately that Keynote's version of each chart is accurate. If it's not, you may need to create the chart again in either Keynote or Numbers.

Note

One of PowerPoint's most useful charting features is being able to link a chart to its underlying source data in an Excel workbook; this way, when the source data changes, you can update the chart automatically to show the latest data. When you open a PowerPoint presentation in Keynote, you lose any links back to an Excel data source. The only workaround is to open the Excel workbook in Numbers, save it as a Numbers document, and recreate the chart from there.

- **Shockwave and HTML links.** Links to Shockwave files or HTML files may disappear. You'll simply need to replace these items once you've saved the presentation in Keynote.

- **Font substitutions.** If the PowerPoint presentation includes fonts that aren't available on your Mac, Keynote substitutes them. Unless any of the fonts in the PowerPoint presentation are so unique that you can't possibly replace them, this substitution has little impact.

Note

See Chapter 12 for instructions on how to export a Keynote presentation in PowerPoint format.

Customizing Slides

Once you've created some slides, you'll often need to move them into a different order. You may also want to arrange them into groups so that you can manipulate them easily. This section shows you how to reorder and group slides; how to add presenter notes; how to set the presentation to skip a slide when you don't want to show it; how to add comments to slides; and how to apply a different theme, master, or layout to a slide to make it look the way you want.

Rearranging and grouping slides

After you create all (or most) of a presentation's slides, you'll often need to shuffle them into the right order.

When you need to move only a single slide, you can drag it up or down the Navigator or the Outline easily. But when you need to move multiple slides, if you switch to light table view, it's much easier to drag one or more selected slides to a new position.

Keynote also lets you organize slides into groups. You can then handle a group of slides as a single unit or collapse the slides you don't want to see. The slide that controls the group is called the *parent* slide; the slides it controls are called *child slides* or *children*.

Here's how to organize slides into groups:

1. **Open the slide navigator if it's hidden.** Click on the View pop-up menu on the toolbar or the View menu on the menu bar, and then choose Navigator.

2. **In the slide navigator, click the slide you want to make a child slide of the slide before it.** For example, if you want to make Slide 2 a child of Slide 1, click Slide 2.

3. **Press Tab or drag the slide to the right until Keynote shows a blue triangle above it, indicating the level to which it will be indented.** Keynote displays a gray disclosure triangle next to the parent slide to indicate that it has one or more children.

4. **Click and drag other slides as needed.** You can indent one child slide under another child slide. Figure 10.10 shows an example of a parent slide with a child slide.

5. **To remove a child slide from a parent, click it and drag it back to the left until Keynote shows a blue triangle at the right indent level.** You can also click the slide and then press Shift+Tab once for each indent you want to remove.

When you've grouped your slides into parents and children, you can click a parent's disclosure triangle to hide its children (if they were displayed) or display them (if they were hidden). By hiding the children you don't want to see (every parent's dream), you can collapse the slides in the slide navigator so that you see the key points of a presentation.

Adding presenter notes

Some people can deliver a word-perfect presentation straight from the slides, but most of us could use some guidance at various times. To help you keep your presentation straight and hit all your talking points, Keynote lets you add presenter notes to each slide. These notes appear on your presenter display only; they're not visible on the monitor the audience is watching.

To add presenter notes to a slide, display the Presenter Notes area by clicking on the View pop-up menu or the View menu on the menu bar and choosing Show Presenter Notes. Keynote displays the Presenter Notes area at the bottom of the window. You can make the area larger or smaller by dragging the sizing handle at the right end of the area up or down.

When you've displayed the Presenter Notes area, simply type the text you need, or paste it in from another document. You can format

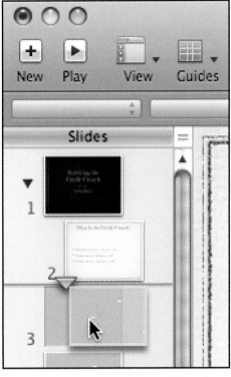

10.10 In the slide navigator, drag a slide to the right to make it a child of the parent slide above it. In this example, Slide 3 becomes a child of Slide 1, which is already the parent of Slide 2.

the text as needed. For example, you can apply boldface to the key points that you want to emphasize.

Setting a slide to be skipped

When you don't need to play a particular slide in a presentation, but you don't want to remove it from the presentation, you can set Keynote to skip it.

You can set skipping on a slide in these ways:

- **Any view.** Click the slide and choose Slide ⇨ Skip Slide.
- **Navigator.** Ctrl+click or right-click the slide and choose Skip Slide.

Solving Font Issues When Transferring Presentations

When you save a Keynote presentation, you can copy the audio files, movie files, and theme images into the presentation file to make sure that these items will be available for the presentation no matter which Mac you open it on.

But you can't include the fonts used in the presentation — so some (or all) of the fonts you need for the presentation may not be available on another Mac.

(By contrast, some versions of Microsoft PowerPoint let you embed all the fonts used in a presentation in the presentation file, making sure that they're available no matter which computer you use to open the presentation.)

There are three main solutions to this problem:

- **Use only fonts you know are installed on the other Mac.** This may sound limiting, but most Macs include such a good variety of fonts that it isn't a hardship for many presentations.

- **Copy the fonts to the other Mac.** If you know which Mac you'll use for the presentation, and there are no licensing issues with copying the fonts, installing them on the other Mac will enable it to display the presentation in its full glory.

- **Substitute the missing fonts with similar ones.** For all but highly specialized fonts, you can usually find a close-enough match among the fonts on the other Mac. Unless your audience is packed with graphic designers, you can be pretty sure they won't notice.

● **Light table view.** Click the slide, then right-click it and choose Skip Slide.

A skipped slide appears in the slide navigator or on the light table as a collapsed gray rectangle; when you select the slide, the rectangle turns yellow, as shown in figure 10.11. In the Outline, the slide also appears collapsed, and has no number next to it.

To remove skipping, repeat the action but choose Don't Skip Slide instead.

Adding comments to slides

Keynote's comments are a handy way to provide input on slides without changing their text. For example, if a colleague asks you to review a presentation, you can use comments to give your suggestions without directly changing the slides in the presentation.

To add a comment to a slide, click the slide, and then click the Comment button on the toolbar or choose Insert ⇨ Comment. You can then type or paste the text of the comment and drag it to a suitable position.

Applying a different theme, master, or layout to a slide

To make your slides appear exactly the way you need them to, you'll often need to change the theme, master, or layout for one or more slides.

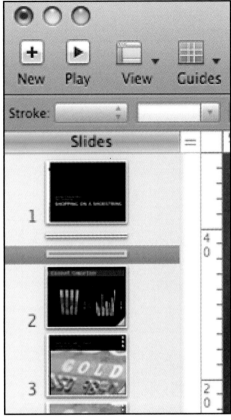

10.11 Keynote displays slides set to be skipped as collapsed rectangles. This example shows two skipped slides between Slide 1 and Slide 2.

Usually, the theme is the best place to start. Click the slide you want to affect, click on the Themes pop-up menu on the toolbar, and then choose the theme you want. If you want to use the Theme Chooser dialog box, choose Theme Chooser at the top of the menu to open the dialog box, which lets you scroll through the various masters contained in each theme.

After you apply the theme, give the slide the master that's best suited for it. With the slide still selected, click on the Masters pop-up menu on the toolbar and choose the master.

After you've applied the master you want to use for a slide, you can adjust the slide's layout as necessary. For example, you can reposition a placeholder on the slide, delete the placeholder, or add another object to the slide. For a smaller change, you can change the font or font size used for one of the text placeholders on the slide.

Importing an Outline from a Document

As you've seen in this chapter, Keynote's Outline pane is great for quickly laying out or adjusting the overall structure of a presentation. When you're creating a presentation from scratch, you can open the Outline pane and type a stream of slide titles and bullet points into it, and then rearrange them as necessary.

But what you'll often need to do is create a presentation from an outline that you already have in a document. If you've created an outline in Pages, using the Heading styles built into many of the templates, you can copy the text from Pages and paste it into the Outline pane in Keynote. Keynote recognizes the styles, so the Heading 1 paragraphs in Pages become the slide titles; the Heading 2 paragraphs become the first-level bullet points; the Heading 3 paragraphs become the second-level bullet points; and so on.

You may find some of the Pages paragraphs come out at the wrong heading levels in Keynote. If this happens, you can fix the problem quickly. To indent a paragraph, demoting it to the next level, select it or click anywhere in it and press Tab. (The insertion point doesn't have to be at the begin-ning of the paragraph.) To reduce or remove a paragraph's indent, promoting it to the next level, select it or click anywhere in it and press Shift+Tab.

If your outline is in a Word document that uses Word's heading styles (which typically use the same names as Pages' styles: Heading 1, Heading 2, and so on), you may be disappointed to find that pasting the Word text into Keynote's Outline pane doesn't produce an outline. Instead, all the headings go to the same level.

Here's the workaround for getting a Word outline into Keynote:

1. **Open the Word document in Pages.** Pages imports the Word styles but changes their names to Heading 1 A, Heading 2 A, and so on so that they don't conflict with its built-in heading styles.

2. **Open the Styles drawer.** Click the Styles Drawer button on the Format bar, press ⌘+Shift+T, or choose Show Styles Drawer from either the View pop-up menu on the toolbar or the View menu on the menu bar.

3. **Switch the Word styles to Pages styles like this:**

   1. In the Styles drawer, move the mouse pointer over the Heading 1 A style.

   2. Click the disclosure triangle, and choose Select All Uses of Heading 1 A.

   3. Click the Heading 1 style to apply that style instead.

   4. Repeat these steps for Heading 2 A, Heading 3 A, and any other heading styles.

4. **Copy the text from Pages, and then paste it into the Outline pane in Keynote.** When you do this, Keynote picks up the heading levels and uses them for the slides. As before, you may need to make minor adjustments if some headings come out at the wrong levels.

5. **Close the Pages document.** If you want to keep it, choose File ➪ Save and save it in a convenient folder. Otherwise, simply close it without saving changes.

Note

If you need to type an actual tab in the Outline pane, press Option+Tab.

How Do I Make My Presentations Lively and Compelling?

All text and no flash make for a dull and forgettable presentation. To keep your audience focused on your presentation and to deliver your message successfully, you'll want to choose a suitable design for your presentation, and plan its content carefully to give it impact. Then add movement, color, and depth to your presentation with movies, audio, and hyperlinks; bring it to life with animations; or give the presentation its own voice by recording narration in sync with the slides. To create powerful presentations of your own quickly and slickly, you'll want to develop your own slide masters and save them in custom themes.

Choosing the Best Theme for the Presentation

The first essential for making your presentation work effectively is choosing a suitable theme for it. Just as you wouldn't write a formal request for a hefty salary raise on a novelty card, you normally wouldn't use a highly colored or super-casual theme for a serious business presentation; you probably would use one of the sober, professional-looking themes instead.

Conversely, if you're giving a high-tempo presentation to kids at your local school or club about a fun topic, you'll be more engaging with a lively and attractive theme than with a strait-laced or pin-striped one.

This may seem to go without saying, but if you sit through a few dozen presentations, you'll often notice glaring mismatches between the theme and the content. Sometimes, this is because the person who created the presentation was so involved in the content that she picked a theme without considering if it would complement the topic. Other times, the person simply doesn't "see" the theme — at least not until the finished slides appear on screen, and the chasm between the look and the content is plain for all to see.

Because the wrong theme can ruin an otherwise great presentation, choose your theme carefully, and be prepared to change it as you develop the presentation. Before showing the presentation to its intended audience, show it to some colleagues who will give you honest feedback. And if they recommend you change the theme, try to look at the presentation with an open mind — because they're probably right.

Giving Your Presentation Impact

Keynote provides plenty of different ways to give your presentations impact. Apart from displaying text in exactly the right font, size, and color on your slides, you can:

- Add images to illustrate points or provide visual interest (covered in Chapter 1).
- Add movies to show action or explain processes in detail.
- Add a soundtrack to give a presentation atmosphere or context.
- Use build effects to reveal, remove, or animate the objects on a slide.
- Use charts and tables to present complex data clearly and emphatically.
- Use transitions to smooth or emphasize the changeover from one slide to another.

All these tools can be useful for creating powerful and effective presentations, but normally, you won't need to use all of them all the time. Resist the temptation to use these tools simply because they're available. Instead, assess realistically what each slide needs in order to present its content clearly and compellingly, apply only those features, and check the slides carefully to make sure you've achieved the effect you were aiming for.

Adding Movies and Audio

To enhance a presentation with sound, you can give it a soundtrack that plays throughout the slide show, add recorded narration synchronized with the slide show, or simply put sounds on individual slides that you can play either automatically or when you click them.

To add visual interest or information, you can add a movie to a slide and choose which part of it to play.

Putting a sound or movie on a slide

To put a sound or movie on a slide, simply open the Media Browser by clicking the Media button on the toolbar, choose the sound or movie, and then drag it to the slide. For example, to add one of your GarageBand compositions that has an iLife preview (a version of the song that you can preview in the Media Browser), click the Audio tab, expand the GarageBand item, and then drag the song to the slide.

Note

You can add a sound or movie by dragging its file from the Finder or by choosing Insert ⇨ Choose.

A sound file appears on the slide as a speaker icon, which you can resize as needed. When you're building the presentation, you can double-click the icon to start playing the sound. For example, you can check that the sound's contents are what you thought they were. Click the speaker icon again when you want to stop playback.

A movie file appears on the slide as the first frame or poster frame from the movie. You can resize the movie as needed, or use the Graphic Inspector to apply a frame to it. Double-click the movie frame to start it playing while you're building the presentation; click it again to stop playback.

After you place the sound file or movie file on the slide, you can choose playback options in the QuickTime Inspector as discussed in Chapter 1. For example, you can choose to start playback after the beginning of the audio or movie file, set the volume, and decide whether to loop the sound or movie.

Genius For a movie, make sure you set the poster frame — the frame you want to display on the slide until the movie starts playing — in the QuickTime Inspector. Your slide will often benefit from having a more dramatic poster frame to start your movie than the very first frame.

Keynote automatically starts the movie or audio file playing as soon as you display the slide or the build (a group of animations) that contains the file. If you want to play the movie file or audio file manually at your convenience, select the Start movie on click check box in the QuickTime Inspector. (Despite its name, this setting works for audio files as well as movies.)

The main types of movies and sound files you can add to Keynote presentations include:

- **QuickTime Movies.** These use the MOV file extension.
- **MPEG-4 movies.** These use the MPEG-4 file extension.
- **Advanced Audio Coding audio files.** These use the AAC file extension.
- **MP3 audio files.** These use the MP3 file extension.
- **AIFF audio files.** These use the AIFF file extension.

Note When you add movie files or audio files to a presentation, you need to make sure that you save the files in the presentation so that they will be included when you transfer the presentation file to another Mac. To include the movie files and audio files, select the Copy audio and movies into document check box in the General pane of the Preferences window or in the Advanced Options section of the Save As dialog box.

Adding a soundtrack to a presentation

The second option for adding audio is to give the entire presentation a soundtrack: the audio starts playing at the beginning of the slide show and keeps playing until the slide show ends or the audio soundtrack ends. If the audio ends first, you can make it loop until the slide show ends.

Here's how to add a soundtrack:

1. **Open the Audio pane of the Document Inspector (see figure 11.1).** Click the Inspector button on the toolbar, click the Document Inspector button, and then click the Audio tab.

2. **Drag the audio you want to the Soundtrack well in the Audio pane one of these ways:**

   - Click iTunes Library to open the Media Browser, and then drag a song or playlist from iTunes.

   - Open a Finder window and drag an audio file from it.

3. **In the pop-up menu, choose Play Once if you want to play the soundtrack once only, or choose Loop if you want to repeat it until the end of the presentation.** Choose Off if you want to switch the soundtrack off temporarily (for example, because you'll give a spoken presentation this time, but don't want to remove the soundtrack from the presentation).

11.1 Use the Audio pane of the Document Inspector to add a soundtrack to a presentation.

4. **Click the Play button, and then drag the Volume slider to set the soundtrack volume.** You may need to tweak the setting when you're ready to give the presentation, but it's usually helpful to set the volume roughly right at first.

Adding Hyperlinks

During a presentation, you'll often need to quickly jump to another slide in the same presentation or to open your default Web browser (for example, Safari) to a particular Web page that contains relevant information. You may also want to open another Keynote presentation instantly without dredging around in the Open dialog box.

These same capabilities can come in handy in presentations you distribute online as well. You may also want to let the viewer quickly create an email message; for example, to give the viewer a way to contact your company or organization.

To create such jumps, you can insert hyperlinks in your presentations like this:

1. **Position the insertion point where you want to create the hyperlink.**

2. **Choose Insert ⇨ Text Hyperlink, and then choose from the submenu the type of hyperlink you want to create: Slide, Webpage, Keynote File, or Email Message.** Keynote inserts a standard hyperlink of the type you chose, opens the Inspector window if it's closed, and displays the Hyperlink Inspector (see figure 11.2).

3. **Choose the target for the hyperlink:**

   ● **Slide.** Click the Next slide option button, the Previous slide option button, the First slide option button, the Last slide option button, the Last slide viewed option button, or the Slide option button and choose the slide number in the pop-up menu.

11.2 In the Hyperlink Inspector, you can choose the destination for the hyperlink, which can be a slide, a Web page, another Keynote presentation, or an email address.

   ● **Webpage.** Type or paste the page's address into the URL box. In the Display box, type the descriptive text you want to display on the slide. For short or widely known URLs, you may want to display the URL itself.

   ● **Keynote Slideshow.** Keynote automatically displays the Open dialog box so that you can pick the presentation file. To change the file, click Choose in the Hyperlink Inspector to display the Open dialog box again. In the Display box, type the descriptive text you want to display on the slide (for example, the presentation's name).

   ● **Email Message.** Type or paste the email address in the To box. Type the default subject line in the Subject box. (The user will be able to change this in his email application.) In the Display box, type the descriptive text you want to display on the slide (for example, Contact Us).

4. **Make sure the Enable as a hyperlink check box is selected.** This makes the hyperlink live, so that Keynote performs the jump when you click the hyperlink.

Genius To change a hyperlink you've inserted on a slide, simply open the Hyperlink Inspector. To delete a hyperlink from a slide, select the hyperlink, and then press Delete.

Adding Animation Builds to Slides

There's nothing like animation to grab and hold your audience's attention, and Keynote lets you animate your slides several ways:

- **Object build.** An object build lets you make an object (such as an image or a table) automatically appear on a slide or vanish from the slide.

- **Object action.** An object action lets you animate an object on a slide. For example, you can animate a chart so that it grows in size or animate a shape so that it spins around.

- **Transition.** A transition animates the switchover from one slide to the next.

Genius If you find that a favorite old animation has disappeared from Keynote, choose Keynote ⇨ Preferences, and then click the General button. In the Animations area, select the Include obsolete animations in choices check box, and then close the Preferences window.

Usually, you'll want to start with object builds and object actions. Once you've applied the builds and actions each slide needs, you can add transitions between the slides.

How object builds work

Keynote lets you use four different kinds of object builds:

- **Build In.** These effects make objects move onto a slide or appear on it.

- **Build Out.** These effects make objects move off a slide or simply disappear from it.

- **Action Build.** These effects animate the objects on a slide (without moving the objects onto the slide or off it, or making them appear or disappear).

- **Smart Builds.** These effects let you use predefined ways of animating an image on a slide.

Revealing objects with Build In effects

You'll probably want to start your work with builds by creating Build In effects that gradually reveal the objects on the slide. A Build In effect lets you focus the audience's attention on a single object, and then on each item you reveal in succession, enabling you to keep future objects up your virtual sleeve until you're ready to introduce them.

Here's how to create a Build In effect:

1. **Create the slide by clicking the New button on the toolbar.** Apply a theme to the slide, choose the best master for it, and then add text and media items. Figure 11.3 shows a slide that contains a title, three bullet points with text, and an image.

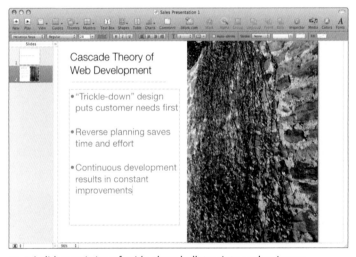

11.3 A slide consisting of a title, three bullet points, and an image.

2. **Open the Build In pane of the Build Inspector.** Click the Inspector button on the toolbar, click the Build Inspector button (the third button from the left), and then click the Build In tab. Figure 11.4 shows the Build In pane with choices selected for the example slide.

3. **On the slide, click the first object you want to reveal.** You can change the order later if you need to.

4. In the Build In pane, set the effect for the object like this:

1. Open the Effect pop-up menu and choose the effect you want. This example uses the Fly In effect.

2. If the pop-up menu to the right of the Effect pop-up menu is available, choose how to treat the object: By Object (which gives you the whole object at once), By Letter, or By Word.

3. If the Direction pop-up menu is available, choose the direction in which to apply the effect — for example, Left to Right, Up, or Random.

4. If you need to change the order of the objects you're animating, open the Order pop-up menu and choose the appropriate number. For example, choose 2 to reveal this object second.

5. If the options in the Delivery area are available, choose how the effect will deliver the object. The Delivery pop-up menu for bullet points lets you choose from All at Once, By Bullet, By Bullet Group (to get a bullet and any subordinate bullets it has), and By Highlighted Bullet. Set the number of seconds for the delivery in the Duration box, and use the Build from controls to set the order — for example, Build from First to Last, or Build from 2 to 5.

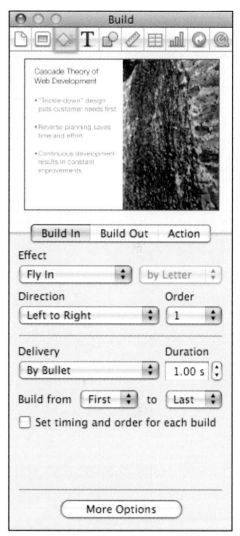

11.4 Use the Build In pane of the Build Inspector to gradually reveal the objects on a slide. The preview at the top shows you the effect your current settings will produce.

Note After each change you make in the Build Inspector, Keynote shows a preview of the effect in its current form. To play the effect again, simply click the preview.

6. If you want to set custom timing for each bullet, reveal the bullets out of order, or do both, select the Set timing and order for each build check box and make changes in the Build Order drawer that the Build Inspector reveals (see figure 11.5). Click the object you want to change, choose an option in the Start Build pop-up menu (for example, Automatically after build 2), and then set the number of seconds to wait in the Delay box. Click Close Drawer when you're ready to close the Build Order drawer.

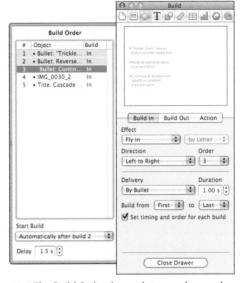

11.5 The Build Order drawer lets you change the order in which you reveal bullets and other objects.

Test the slide full-screen by clicking the Play button on the toolbar to make sure the effect looks the way you intended.

Removing objects with Build Out effects

When you want to remove objects from a slide, use a Build Out effect. This is great for removing items you no longer need and leaving the audience with only your takeaway point to look at.

To create a Build Out effect, you use the same techniques as for a Build In effect (as described previously), but you work in the Build Out pane instead.

Genius If you want to apply the same build effect to several slides of the same type, create the build effect on the slide master rather than on the slides. See the end of this chapter for instructions on creating your own slide masters.

Animating objects with Action Build effects

When you need to animate an object on a slide, use the Action pane of the Build Inspector to create an Action Build. Keynote provides a good variety of actions, including:

- **Move.** This effect moves an object from one place on a slide to another.

- **Scale.** This effect increases or decreases the size of an object.

- **Rotate.** This effect makes an object spin around.

- **Opacity.** This effect lets you make an object fade in or out.

How you create an Action Build depends on the action you choose, as different actions need different settings. Here's an example of creating an Action Build using the Move effect:

1. **Open the Action pane of the Build Inspector.** Click the Inspector button on the toolbar, click the Build Inspector button, and then click the Action tab. Figure 11.6 shows the Action pane with choices made for a slide.

2. **On the slide, click the object you want to move.**

3. **In the Effect pop-up menu, choose the Move effect.** Keynote displays a red move path starting at the object's center.

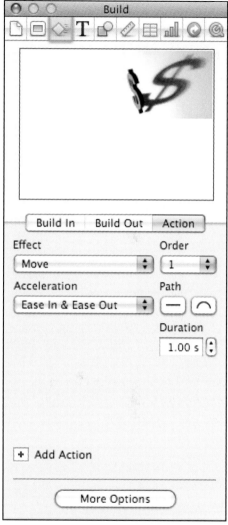

11.6 The Action pane enables you to set up custom movement for an object, make it grow or shrink, or make it rotate — among other actions.

297

4. **Click and drag the move path's end point to where you want it.** Figure 11.7 shows an example.

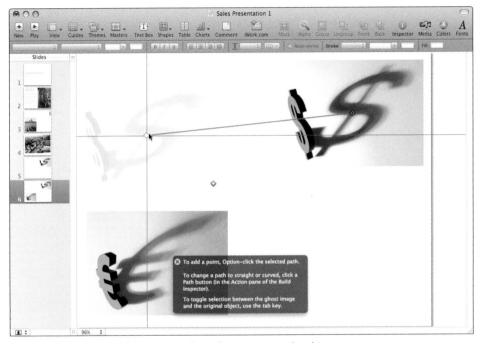

11.7 Drag the endpoint of the move path to where you want the object to go.

5. **If you need to customize the path, do so in one of these ways:**

   - To create a curved path, click the Curved button in the Path area, and then drag the central point to change the curve. To create a straight path, click the Straight button.

   - To add a change of direction, Option+click the move path to place a point, and then click and drag the point to where you want it.

6. **In the Acceleration pop-up menu, choose the type of acceleration to use.** None gives steady movement, Ease In starts off gradually, Ease Out ends gradually, and Ease In & Ease Out both starts and ends gradually.

7. **In the Duration box, set the number of seconds for the action to last.**

8. **If you need to add another path to the object's movement, click the Add Action button near the bottom of the Action pane.** Keynote adds another move path, which you can then adjust as described previously.

Note Keynote displays a red badge on the lower-right corner of an object to indicate that it has an Action Build applied to it.

Using Smart Builds

Keynote's Smart Builds give you an easy way to show a series of images on a slide with animated transitions between each pair of slides. Each Smart Build comes configured with transitions that you can either use as they are or customize to suit the needs of your presentation.

Here's how to add a Smart Build to a slide:

1. **Choose Insert ⇨ Smart Build, and then click the effect you want from the Smart Build submenu.** Your choices are Dissolve, Flip, Grid, Push, Revolve, Shuffle, Spin, Spinning Cube, Swap, Thumb Through, and Turntable. This example uses Thumb Through. Keynote displays a blue Smart Build placeholder and a dialog box for the Smart Build — for example, the Thumb Through dialog box — and opens the Media Browser with the Photos pane showing.

2. **Resize and reposition the Smart Build placeholder as needed.** Click and drag the placeholder's handles to make it take up as much of the slide as you want to devote to the images. To reposition the placeholder, click in it and drag.

3. **Drag photos from the Photos pane to the wells in the Smart Build's dialog box.** Figure 11.8 shows an example with four photos added to the Thumb Through dialog box.

4. **If you want all the images to appear at the same size, select the Scale images to same size check box.** Deselect this check box if you want each image to appear at its original size.

5. **Click the Play button on the toolbar to test the Smart Build full screen.** If you want, use the Action pane of the Build Inspector to customize the Smart Build's transitions.

Note Keynote displays a purple badge on the lower-right corner of an object to indicate that it has a Smart Build applied to it.

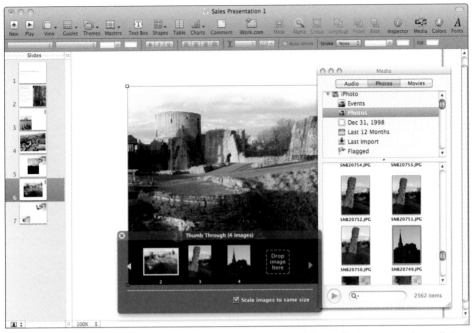

11.8 To create a Smart Build, drag photos from the Photos pane of the Media Browser to the wells in the Smart Build dialog box (here, the Thumb Through dialog box).

Using builds with tables and charts

When you're presenting complex data using a table or a chart, you can apply a custom build on the Build In pane or Build Out pane of the Build Inspector to make the information easier to read. Table 11.1 lists the delivery options you can use for tables.

Table 11.1 Keynote's Delivery Options for Tables

Delivery Option	What It Does
All At Once	Reveals or removes the table all at once.
By Row	Reveals the table one row at a time.
By Row Content	Reveals the whole table's frame, and then displays the content one row at a time.
By Column	Reveals the table one column at a time.
By Column Content	Reveals the whole table's frame, and then displays the content one column at a time.

Delivery Option	What It Does
By Cell Content	Reveals the whole table's frame, and then displays the content one cell at a time — across the first row cell by cell, then across the second row cell by cell, and so on.
Bottom Up	Reveals the table's last row, second last row, and so on up.
Bottom Up – Content	Reveals the whole table's frame, and then reveals the table's last row, second last row, and so on up.

Similarly, Keynote lets you reveal a chart either all at once or piece by piece. Table 11.2 lists the delivery options you can use for charts.

Table 11.2 Keynote's Delivery Options for Charts

Delivery Option	What It Does
All At Once	Reveals or removes the chart all at once.
Background First	Reveals the chart axes, and then reveals the data elements all at once.
By Series	Reveals the chart axes, and then reveals the data, series by series.
By Element in Series	Reveals the chart axes, and then reveals the data, element by element within a series.
By Set	Reveals the chart axes, and then reveals the data, set by set.
By Element in Set	Reveals the chart axes, and then reveals the data, element by element within a set.

Adding Transitions between Slides

A transition is a visual effect that Keynote plays to animate the switchover from one slide to another. Keynote provides a wide range of transitions that you can use to add visual appeal to your presentations. For example, the Push transition causes the incoming slide to push the outgoing slide out in the direction you choose (for example, from right to left), while the Cube transition uses the same rotating-cube effect as Mac OS X's Fast User Switching feature.

Genius

Be careful not to overdo transitions. They're fun to set up, but don't spend your time on transitions when you should be honing the presentation's content and message. And they can be fun for the audience to watch — up to a point. If your transitions draw more attention than your slides, you probably won't get your point across. For formal presentations, you're usually better off sticking to a single, subtle transition (such as Dissolve) for the whole presentation.

Here's how to add a transition between two slides:

1. **Click the slide after which you want the transition to appear.**

2. **Open the Transition pane of the Slide Inspector (see figure 11.9).** Click the Inspector button on the toolbar, click the Slide button, and then click the Transition tab.

3. **In the Effect pop-up menu, choose the transition you want.** Keynote breaks the transitions down like this:

   - **None.** Select this item to let the slides switch over without any effects.

   - **Recent Effects.** This category lets you quickly choose again one of the effects you've used recently rather than having to plumb the depths of the list.

   - **Magic Move.** This is a special effect you can use for similar slides. See the nearby sidebar for details.

   - **Text Effects.** These effects — Anagram, Shimmer, Sparkle, and Swing — play with the letters in the words. They work best with text-based slides dominated by a few words.

   - **Object Effects.** These effects — Object Push, Object Zoom, Perspective, and Revolve — move all the slide's objects off the first slide and then replace them with the objects from the next slide. These effects work best with slides dominated by objects rather than by text.

11.9 Use the Transition pane of the Slide Inspector to choose a transition effect and set options to make it play the way you want.

- **3D Effects.** These effects — Cube, Doorway, Flip, Flop, Mosaic, Page Flip, Reflection, Revolving Door, Snap, and Twist — move one slide off the screen and the next slide onto the screen with three-dimensional visual effects. These effects may be jerky if your Mac's graphics card is underpowered.

- **2D Effects.** These effects — Dissolve, Fade Through Color, Iris, Move In, Push, Reveal, Scale, and Wipe — move one slide off the screen and the next slide onto the screen with two-dimensional visual effects. These are the most subtle effects and work fine on just about any graphics card.

Note If your Mac is an older one, you may see a category called Effects that can't play on this computer at the bottom of the Effect pop-up menu. This simply means that your Mac's processor doesn't have the horsepower to run this effect. If you are going to give the presentation on a more powerful Mac, you can apply this transition once you've moved the presentation to that Mac.

4. **Watch the preview for the transition.** Keynote automatically plays a preview after you choose the transition and after you adjust any of its settings.

5. **In the Duration box, set the number of seconds you want the transition to run for.** Depending on the contents of the slide, the transition may look better at a faster or slower speed, so it's worth experimenting a bit.

6. **If the Direction pop-up menu is available, choose the direction for the transition.** The choices vary depending on the transition, but these are typical: Left to Right, Right to Left, Top to Bottom, Bottom to Top, and Random.

7. **In the Start Transition pop-up menu, choose when to run the transition.** For a presentation that you will run manually or that someone else will advance through at her own pace, choose On Click. For a presentation that will run itself, choose Automatically and set the time in the Delay box.

8. **Choose other settings for the transition.** The settings vary depending on the transition you've chosen. For example, the Mosaic transition has a Type pop-up menu that lets you choose among Dissolve, Flip, Scale, Spin, and Pop options, and a Size pop-up menu that offers Small, Medium, and Large.

9. **Close the Slide Inspector when you're satisfied with your choices.**

Creating Magic Move Transitions

Keynote has a neat transition called Magic Move that works with two slides that contain one or more of the same objects — for example, an image or a text box. You can't use Magic Move with charts, tables, or movies.

If you want to use Magic Move with just a single object, you can copy it from one of the slides, paste it on the other, and then move it to where you want it to appear. But if you want to use Magic Move with several objects, the easiest thing to do is duplicate the slide.

When you've finished creating the first slide, Ctrl+click or right-click it in the slide navigator and choose Duplicate. Keynote creates a duplicate of the slide and places it after the original. You can then move the objects on the duplicate to their new positions, and then apply the Magic Move transition to the original. When you play the transition, the objects on the first slide move to their new places on the second slide.

Adding Narration

If you're planning to distribute a presentation rather than deliver it in person, you may want to add narration to it. Keynote makes this easy.

Caution Wait until you've finished creating your presentation, including all builds and transitions, before you add narration to it. Otherwise, any adjustments you make to the presentation after adding the narration are likely to foul up the timing.

Here's how to add recorded narration to a presentation:

1. **Set up your microphone.** If you have a quiet location to record in, your Mac's built-in microphone may do the trick. Generally, however, you'll get better results by using a microphone you can position close to your mouth but not directly in your breath stream (blowing on the microphone tends to produce rasps and distortions).

2. **Decide which slides you will record narration for.** In many cases, you'll want to record narration for the entire presentation — in which case, simply start at the first slide and carry on to the end. But you can also record narration for only some slides if you prefer. Just pick the slide you want to start with, and continue to where you want to stop. When you play back the recording, Keynote plays back only those slides.

To select the microphone, choose Apple ⇨ System Preferences, click Sound, and then click the Input button. In the Choose a device for sound input box, click the microphone you want. Speak into the microphone, and then drag the Input volume slider to set the level. Select the Use ambient noise reduction check box if you want Mac OS X to filter out ambient sounds.

3. **Choose File ⇨ Record Slideshow to start the recording.** If you have the Audio pane of the Document Inspector open, you can click the Record button there instead. Keynote starts playing the slide show from the slide you chose and displays a blinking red light in the upper-left corner of the screen to show that it's recording, together with a volume meter that shows the input volume the Mac is picking up from the mike.

4. **Speak your narration into the microphone.** Click the mouse or press Right Arrow when you're ready to display the next slide.

Note If you need to take a break during the recording, you can pause the recording by clicking the recording indicator. You can also press B (the black-screen key), W (the white-screen key), or F (freeze) to pause recording. To resume recording, press any key.

5. **To stop recording, press Esc.** Keynote stops the slide show and saves the recording in the presentation file.

Now play back the slide show and listen to the narration. If you need to adjust the playback volume, open the Audio pane of the Document Inspector and drag the Volume slider. You can also play back the narration *without* watching the slide show by clicking the Play button (the button with the Play symbol) in the Slideshow Recording area in the Audio pane.

Adding to your recorded slide show

If you recorded narration for only some of the slides in the slide show, you can add narration to other slides. Click the slide at which you want to start, choose File ⇨ Record Slideshow, and then click the Record & Append button in the Are you sure you want to record from the current slide? dialog box that appears (see figure 11.10).

If you see this dialog box when you want to record the whole slide show again, click the Record From Beginning button.

Are you sure you want to record from the current slide? The new recording will be appended to the existing recording.

To replace the entire recording, click Record from Beginning.

Record From Beginning   Cancel   Record & Append

11.10 When you've recorded narration for only part of a slide show, you can append recorded narration for the other slides.

Rerecording a slide show

If the recording isn't up to scratch or is missing essential information, you can redo it by selecting the starting slide and choosing File ⇨ Record Slideshow (or clicking the Record button in the Slideshow Recording area in the Audio pane of the Document Inspector). Keynote displays the Are you sure you want to rerecord your slideshow? dialog box (see figure 11.11) to make sure you know you're overwriting the previous recording. Just click the Record Again button, and then say your piece.

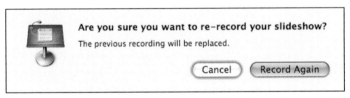

Are you sure you want to re-record your slideshow?

The previous recording will be replaced.

Cancel   Record Again

11.11 To rerecord narration, you must acknowledge that you're overwriting the existing recording.

Note   If you simply want to get rid of the recorded narration without replacing it, choose File ⇨ Clear Recording or click the Clear button in the Slideshow Recording area in the Audio pane of the Document Inspector. Click Clear in the confirmation dialog box that Keynote displays.

After you record narration in a slide show and save it, when you start editing the slide show again, Keynote double-checks to make sure you want to do this (see figure 11.12) because it may put the recording out of sync with the slide show. Click Edit Show if you want to proceed. Alternatively, click Save As to save a copy of the presentation that you can edit without affecting the original.

Are you sure you want to edit your slideshow? Editing your slides might make the recording out of sync with the slideshow.

To save a copy of this slideshow to edit, click Save As.

(Save As...)   (Cancel) (Edit Show)

11.12 Keynote double-checks to make sure you want to edit a presentation that includes recorded narration.

Genius

When you want to play a slide show without recorded narration, you don't need to remove the narration — you can simply suppress it. To do so, open the Presentation pop-up menu in the Document pane of the Document Inspector and choose Normal instead of Recorded.

Creating Your Own Slide Masters

As you've seen, each of Keynote's themes includes various master slides for different needs — title slides, slides with horizontal or vertical photos, slides with charts, slides with tables, and so on.

If the masters in the themes you use meet your needs, you're all set. But if you find you need to customize the slides you create from the masters, you can save time by creating your own masters that contain exactly what you need.

Genius

If another theme or presentation contains a suitable slide or master, you can simply import it into the presentation. Open a presentation that contains the slide or master you want, and arrange the Keynote windows so that you can see both the source window and the destination window. Then drag the slide from the source to the slide navigator in the destination window, or drag the master from the source to the master slide navigator in the destination window.

Here's how to change a master slide:

1. **Open the master slide navigator.** Choose Show Master Slides from either the View pop-up menu on the toolbar or the View menu on the menu bar. Or simply drag down the handle on the title bar of the slide navigator to expose the master slide navigator.

Genius

2. **Click the master slide you want to change.** Keynote displays its contents, together with the master gridlines and alignment guides (see figure 11.13).

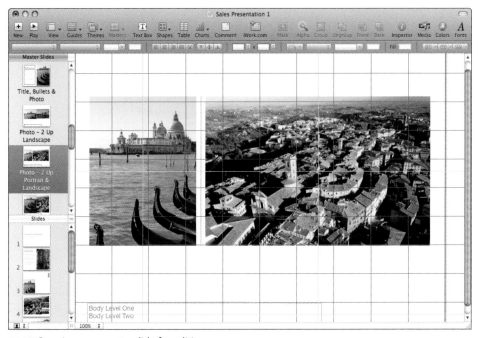

11.13 Opening up a master slide for editing.

Note

If you see too many or too few master gridlines, open the Preferences window and adjust them. Choose Keynote ➪ Preferences to open the Preferences window, then click the Rulers tab to display the Rulers pane. Adjust the Show horizontal gridlines every setting to show fewer or more horizontal gridlines; adjust the Show vertical gridlines every setting to control the number of vertical gridlines displayed.

3. **Open the Appearance pane of the Master Slide Inspector (see figure 11.14).** Click the Inspector button on the toolbar, and then click the Master Slide Inspector button (the second button from the left). Click the Appearance tab.

4. **In the Layout area, select the check box for each of the items you want to include: Title, Body, Object Placeholder, or Slide Number.** Drag each item to where you want to position it, and resize it as needed.

5. **Select the Allow objects on slide to layer with master check box if you want to be able to interleave objects you place on the slide with objects in the master.** For example, you can place an object in the slide partly over an object in the master, but then send the object in the slide behind the object in the master.

6. **If you want to create a media place-holder on the slide, select the Object Placeholder check box in the Layout area.** Click the placeholder, resize and reposition it as needed, and then select the Define as Media Placeholder check box. Type a name or description for the media placeholder in the Tag box to make clear to anyone using the presentation what type of object the placeholder is for. You can also drag a placeholder image to the placeholder if you want.

11.14 In the Appearance pane of the Master Slide Inspector, choose which objects to include, and set up any background fill, gradient, or picture the master needs.

Note

Changes you make to a master slide in a presentation carry through to each of that presentation's slides that are based on the master.

7. **If the master slide needs a fill, gradient fill, image fill, or tinted image fill, use the controls in the Background area of the Appearance pane to set it up.** See Chapter 1 for detailed instructions.

8. **Add any text or images needed to the background of the master slide.** For example, drag in a theme graphic or a company logo from the Media Browser, or place a text box and enter text.

Genius

After you add a background image or text placeholder to a master slide, it's a good idea to lock it in place to prevent anyone from moving it. Click the image or object, and then choose Arrange ⇨ Lock. (If necessary, you can unlock it by selecting it and choosing Arrange ⇨ Unlock.)

9. **If you want to set a default transition for the master slide, click the Transition button in the Master Slide Inspector to display the Transition pane.** You can then choose the effect, duration, direction, and other options as described earlier in this chapter.

10. **If you want to set up default builds for the master slide, open the Build Inspector and set up the builds as described earlier in this chapter.**

When you've finished customizing the master slide, save changes to the presentation, and then customize the other master slides that need changes.

Creating a Custom Theme

Creating your own masters can save you a great deal of time, but once you've created your masters, you'll probably want to take the next step as well — to create one or more of your own custom themes. This is the best way to keep your custom masters and custom content together in a format you can reuse in moments.

You can create a custom theme one of these ways:

- **Customize an existing theme.** If one of Keynote's themes can provide a suitable starting point, this is usually the quickest way to create a custom theme.

- **Create a new theme from scratch.** Create a new presentation by pressing ⌘+N or choosing File ⇨ New. Delete all the master slides but one, and then start creating master slides from that master.

When you've set up the theme to your liking, choose File ➪ Save Theme, type a descriptive name for the theme, and then click Save.

You can then start a new presentation based on your custom theme by choosing File ➪ New from Theme Chooser, clicking your theme at the bottom of the Theme Chooser window, and then clicking Choose.

What is the Best Way to Give My Presentation?

The traditional way to deliver a presentation is in person, using a computer and perhaps an external projector, and Keynote gives you a full suite of tools to do this, as well as a way to provide your audience with informative handouts. Keynote also lets you create a presentation that plays automatically, which is great for a kiosk at a trade show or similar event, or for sharing a presentation in a host of other ways. For example, you may want to produce a PowerPoint slide show or a QuickTime movie from your Keynote presentation; share it via the Internet as Web pages, PDF files, or image files; turn a slide show into a podcast; or simply publish it directly from Keynote to YouTube.

Setting Preferences for the Presentation

Before you give a presentation, you'll want to make sure the settings in the Slideshow pane, Presenter Display pane, and Remote pane in Keynote's Preferences window are set the way you need them. Choose Keynote ➪ Preferences or press ⌘+, (comma) to open the Preferences window, and then click the Slideshow tab to display the Slideshow pane.

Setting Slideshow pane preferences

The Slideshow pane in Keynote's Preferences (see figure 12.1) lets you control how the application appears and acts while you're giving the presentation.

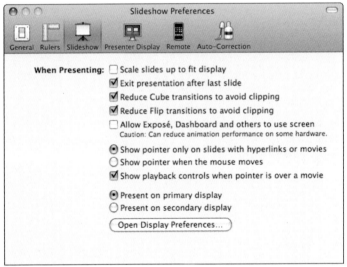

12.1 Use the Slideshow pane in the Preferences window to control how Keynote appears and behaves while you're delivering a presentation.

Choosing slide options

Select the Scale slides up to fit display check box if you're using slides that are lower resolution than the display you're using. For example, if your slides are 800 × 600 resolution and the display is 1024 × 768 resolution, you may want to scale them up so that they fill the entire screen. The disadvantage to scaling the slides up is that video quality will usually suffer a bit.

Select the Exit presentation after last slide check box if you want Keynote to automatically close the presentation after the last slide finishes. Deselect this check box if you prefer to keep the final slide on-screen as you finish the presentation. (For example, you might display talking points or your company's contact information.)

Select the Reduce Cube transitions to avoid clipping check box and the Reduce Flip transitions to avoid clipping check box if you use either of these transitions on a presentation you're scaling up. In this case, *clipping* means that part of the transition gets cut off because the slide has changed size.

Choosing whether to enable Exposé and Dashboard

Normally, you won't want to use Mac OS X's navigation features such as Exposé, Spaces, and Dashboard during a presentation, so Keynote turns them off while you're running a presentation. If you do want to use them, select the Allow Exposé, Dashboard and others to use screen check box.

Choosing preferences for the mouse pointer

Usually, when giving a slide show, you won't want the mouse to appear on-screen for the most part because it can be distracting to the audience. On the other hand, it's likely you will want to use the mouse to control the presentation and move from slide to slide.

To help you, Keynote gives you a choice of options for the mouse:

- **Show pointer only on slides with hyperlinks or movies.** Select this option button if you want the mouse pointer to appear on such slides so that you can click the hyperlinks or play the movies.

- **Show pointer when the mouse moves.** Select this option button if you want Keynote to keep the pointer hidden until you deliberately move the mouse. (Clicks don't count.)

Whichever option you prefer, you can select or deselect the Show playback controls when pointer is over a movie check box to make Keynote display the playback controls for movies when you move the pointer over them.

Choosing which display to present on

To choose which display to show the presentation on, select the Present on primary display option button or the Present on secondary display option button. For example, if you're using a MacBook, the primary display is the built-in screen, and the secondary display is an external display or projector you've connected to the MacBook. In this case, you'll normally present on the secondary display and use the primary display for viewing your *presenter display* — your view of the current slide, your notes for it, the upcoming slide, and so on.

Setting Presenter Display pane preferences

The Presenter Display pane in the Preferences window (see figure 12.2) lets you customize the information that Keynote displays for you, the presenter, rather than for the audience watching the presentation. By tweaking these preferences, you can make sure you see exactly the information you need to deliver your presentation with the greatest impact — and as easily as possible.

Genius

While many of Keynote's preferences are "set-and-forget" ones that work for any presentation, the preferences in the Presenter Display pane are different. Chances are that you'll often need to change these preferences depending on the type of presentation you're giving and the audience to which you're delivering it.

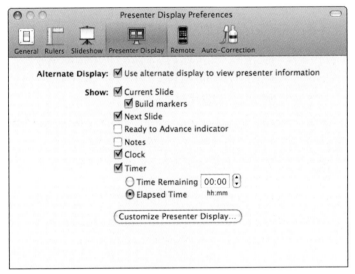

12.2 The preferences in the Presenter Display pane let you set Keynote to handle the presentation exactly the way you want it.

Setting Alternate Display preferences

There's only one preference in the Alternate Display area, but it's a key one. Select the Use alternate display to view presenter information check box if you want Keynote to put your presenter information on a different display than the one the presentation is on. For example, you'll often have your presenter information on your MacBook and the presentation on a projector or external display.

Setting Show preferences

In the Show area of Presenter Display Preferences, you can choose the following preferences:

- **Current Slide.** Select this check box if you want to have the current slide appear in the presenter display. Seeing this is usually handy as long as your screen has enough space to display the slide along with all the other information you need; if the screen is too small, you may choose to forego the slide, especially if you can easily see what the audience is viewing.

- **Build markers.** This check box is available only if you select the Current Slide check box. You can then select the Build markers check box to display the build markers that represent the animations applied to the current slide. Seeing the build markers can be helpful for keeping what you're saying synchronized with the different builds you're revealing.

- **Next Slide.** Select this check box if you want your display to show the next slide. Usually, seeing the next slide is helpful for keeping tabs on what's coming up and, thus, keeping your presentation firmly on track. If you don't have enough space on-screen, you may prefer to sacrifice the next slide so that you have more space for the current slide. The next slide appears at a smaller size than the current slide.

- **Ready to Advance indicator.** Select this check box if you want Keynote to display a red bar across the top of the presenter display to indicate that a slide is still playing. When the slide has finished playing, and Keynote is ready to advance to the next slide, the bar becomes green. While this signal is often helpful, especially when you're delivering a presentation that someone else has created, it can also be distracting.

- **Notes.** Select this check box to display your notes across the bottom of the presenter display. Depending on how well you know your material, seeing the notes can be essential, helpful, or merely superfluous.

- **Clock.** Select this check box to display the clock with the current time so that you can see how you're doing.

- **Timer.** Select this check box if you want to see a timer on the presenter display. You can then select the Time Remaining option button and set the number of hours and minutes in the box, giving yourself a countdown, or select the Elapsed Time option button so that you see a straightforward readout of how long you've been presenting for.

Customizing the presenter display

If you want to customize the presenter display further, click the Customize Presenter Display button at the bottom of the Presenter Display Preferences pane. Keynote displays the Customize Layout screen (see figure 12.3) on which you can choose which items to display, where to display them, and how big they should be. Customize the presenter display as follows:

1. **In the Customize Presenter Display dialog box, select the check box for each item you want to display.** These are the same items that appear in the Show area of the Presenter Display Preferences.

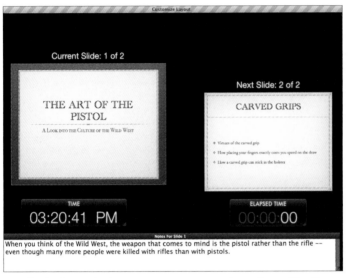

12.3 On the Customize Layout screen, you can resize and reposition the items that appear in the presenter display.

2. **To reposition an item, click it and drag it to where you want it to appear.** For example, you can drag the Elapsed Time readout up toward the top of the display.

3. **To resize an item, drag the handle in its lower-right corner.**

When you've finished customizing the presenter display, click the Done button in the Customize Presenter Display dialog box to return to the Preferences window.

Setting Remote pane preferences

The Remote pane in Keynote's Preferences pane window lets you set up an iPhone or an iPod touch as a remote control for Keynote, so that you can use the device to run your presentations.

If you have an iPhone or iPod touch and you use Keynote, the Remote application is a great use for the device. If you don't have one, it's a strong argument for your company to buy one for you.

Here's how to set up an iPhone or iPod touch as a remote:

1. **Download the Remote application from the App Store section of the iTunes Store.** Usually, it's easiest to download the application using iTunes and then synchronize it to the device, but if you prefer, you can download it directly to the iPhone or iPod touch and then synchronize it back to iTunes.

2. **On the iPhone's or iPod touch's Home screen, touch the Remote application's icon to display the Keynote Remote screen.**

3. **Touch the Link to Keynote button to display the Settings screen (see figure 12.4).** You may not even have to touch the button: When you haven't yet linked to Keynote, the Remote application automatically displays the Settings screen for you after a moment.

4. **Touch the New Keynote Link button to display the New Link screen.** This shows a four-digit passcode and instructions.

5. **On your Mac, select the Enable iPhone and iPod touch Remotes check box in the Remote pane in the Preferences window (see figure 12.5).**

6. **On your Mac, click the Link button for the iPhone or iPod touch in the list box.** In the Add Remote for iPhone and iPod touch dialog box that appears, type the passcode that the iPhone or iPod touch is displaying. When you enter the passcode correctly, this dialog box closes.

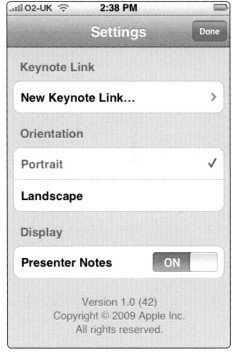

12.4 Touch the New Keynote Link button on the Settings screen for the Remote application to start setting up an iPhone or iPod touch as a remote.

12.5 Setting up an iPhone or iPod touch as a remote for Keynote.

7. **On your Mac, an Unlink button replaces the Link button for the iPhone or iPod touch.** On the iPhone or iPod touch, the New Link screen closes, and your Mac's name appears in the Keynote Link area of the Settings screen.

8. **In the Orientation area, touch Portrait or Landscape to choose the orientation you will use.**

Note
Portrait orientation lets you see presenter notes on the iPhone or iPod touch, but you can see only the current slide. Landscape orientation lets you see both the current slide and the next slide, but not the presenter notes.

9. **If you chose Portrait orientation, choose whether to display presenter notes.** In the Display area, slide the Presenter Notes switch to On if you want to see presenter notes on the iPhone or iPod touch, or to Off if you don't.

You're all set to control a Keynote presentation from the iPhone or iPod touch.

Giving a Live Presentation

When you're gearing up to give a live presentation, you'll want to finalize your presenter notes, rehearse the presentation, and connect your Mac to an external display or projector and make sure it's set up so that you can see what you need.

Finalizing your presenter notes

All too often, when you're getting ready to give a live presentation, you realize that your presenter notes contain nothing like the detail you need — or that some slides don't have any presenter notes at all.

If you have time, take a few minutes to go through the slides patiently and methodically. Check the presenter notes for each slide, rehearse what you will say for the slide, and make sure the two match. Add brief phrases or sentences to the presenter notes area summarizing each point you want to make certain you hit, and drag the paragraphs into the order in which you plan to hit the points.

Genius

You can print out your presenter notes by selecting the Slides With Notes option button in the Print area of the Print dialog box.

Rehearsing the presentation

To help you polish your presentation, Keynote includes a rehearsal view that you can use with a single monitor — a great advantage over having to add an external display in order to practice. The rehearsal view is the same as presenter display except that there's no screen for the audience.

Here's how to rehearse your presentation:

1. **From the menu bar, choose Play ⇨ Rehearse Slideshow.** Keynote switches to rehearsal view and displays the first slide and the slide after it (see figure 12.6).

2. **Practice delivering your presentation, referring to your presenter notes as needed.** Use the Time display and Elapsed Time display (if you chose to show them) to track how long you're taking.

Note

If your presenter notes don't appear in rehearsal view, move the mouse pointer up to the top of the screen to display the presentation toolbar, click the Options pop-up menu, and then select Customize Presenter Display. In the Customize Presenter Display dialog box, select the Notes check box, and then click Done.

12.6 Rehearsal view lets you run through your presentation and practice your timings using only a single screen.

3. **Click to display the next build or slide.** Ctrl+click or right-click to move back a step in the presentation.

4. **To move to other slides, use the slide switcher.** Move the mouse pointer to the top of the screen and click Slides on the presentation toolbar. In the slide switcher (see figure 12.7), double-click the slide you want, or type its number and press Return. Keynote closes the slide switcher for you.

12.7 The slide switcher lets you quickly jump from slide to slide in your presentation.

5. **Press Esc when you want to end the slide show.**

Setting up an external display or projector

When the time comes to deliver the presentation, plan to give yourself plenty of time to set up your Mac with the external display (or displays) or projector you'll use. Setup should be swift and painless, but it's easy to become all thumbs when you're in a rush.

Genius

When scheduling your presentation and your travel arrangements, allow yourself extra time not only for setting up your equipment and troubleshooting unexpected problems but also for taking a practice run right through your presentation without worrying that the audience is about to walk in. Having rehearsal time and a chance to relax and focus can help your presentation immeasurably.

When your audience is small enough to count on the fingers of one hand, presenting on your Mac's own display (for example, a MacBook Pro's built-in screen) is viable. But usually, you'll need to use either an external display or a projector for the audience and your Mac's display for yourself.

Connect the external display or projector to your Mac following the instructions that came with both devices, and then use the Displays preferences like this to tell Mac OS X which display is where:

1. **Open the Displays preferences.** Click the System Preferences icon on the Dock (if it appears there) or choose Apple ➪ System Preferences to open System Preferences. Then click the Displays item in the Hardware area. You'll see a Display preferences window for each display, so you can choose separate settings for each.

2. **On the Display tab for the external display or projector, choose the resolution to use.** For example, choose 1024 × 768 for a projector. Normally, you won't need to change the resolution for your Mac's display.

3. **Click the Arrangement button to display the Arrangement tab (see figure 12.8).**

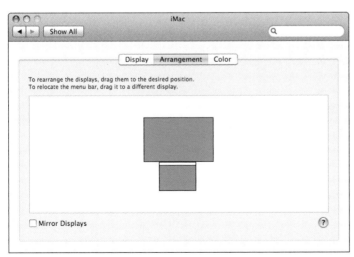

12.8 On the Arrangement tab of the Displays preferences, click and drag the icons for the displays so that they're arranged the same way as the physical displays.

323

4. **Click and drag the icons for the displays to match the way the physical displays are arranged.** For example, if you'll have your MacBook on a desk and will look above it to view the projection screen, position the projection screen's display icon above the MacBook's display icon.

Note

Selecting the Mirror Displays check box on the Arrangement tab in the Displays preferences makes Mac OS X display the same information on each display. This can be useful if you're presenting on two displays (with some of your audience viewing each display), but otherwise you won't normally want to use it because it prevents you from viewing your presenter display.

5. **Close the Displays preferences.** For example, press ⌘+Q.

Checking your presenter display

When you've set up the hardware and arranged the displays, check that your presenter display appears on the display you want. This will help you avoid a heart-stopping moment when you start the presentation in earnest and find yourself greeted with the audience's view, while your presenter notes are magnified for the audience to see.

Choose Play ⇨ Play Slideshow to test the slide show. If you find that the displays are reversed, just press X to switch them over.

If only the mouse is available, move the mouse pointer to the top of the screen so that the presentation toolbar appears, click the Options button, and then choose Swap Displays from the pop-up menu.

Running the presentation

To start the presentation running, click the slide you want to start with, and then click the Play button on the toolbar or choose Play ⇨ Play Slideshow.

To start the presentation running from the first slide, Option+click the Play button on the toolbar.

The first slide appears on the audience's display, and the first two slides appear on the presenter display, along with the presenter display tools you've chosen (for example, presenter notes, the clock, and the timer).

You can then run the presentation using the mouse, the keyboard, an Apple Remote, or an iPhone or iPod touch, as discussed in the next sections.

Controlling the presentation with the mouse

Here's how to control the presentation with the mouse:

- **Display the next build or slide.** Simply click.

- **Display the previous build or slide.** Ctrl+click or right-click.

- **Display the slide switcher.** Move the mouse pointer up to the top of the screen to display the presentation toolbar, and then click Slides. Double-click the slide you want to display.

  Black out the screen. Move the mouse pointer to the top of the screen to display the presentation toolbar, and then click Black. Click anywhere when you want to remove the black screen and make the presentation visible again.

Controlling a presentation using keyboard shortcuts

You can also run the presentation using the keyboard shortcuts shown in Table 12.1.

Table 12.1 Keyboard Shortcuts for Running a Presentation

Action	Keyboard Shortcut
Play the slide show from the first slide	Option+click the Play button
Play the slide show from the current slide	Command+Option+P
Show the first slide	Home
Show the last slide	End
Show the next build	N, Space bar, Return, Page Down, Right Arrow, or Down Arrow
Show the previous build	Shift+Left Arrow, Shift+Page Up, or [
Show the next slide	Shift+Down Arrow, Shift+Page Down, or]
Show the previous slide	P, Delete, Page Up, Left Arrow, Up Arrow, or Shift+Up Arrow
Show the last slide viewed (not necessarily the previous slide)	Z
Hide Keynote, showing the last application used	H
Freeze the presentation at the current slide	F

continued

325

Table 12.1 continued

Action	Keyboard Shortcut
Show a black screen (pausing the presentation)	B
Show a white screen (pausing the presentation)	W
Resume the presentation from frozen, black screen, or white screen	(Press any key)
Show or hide the mouse pointer	C
Display the slide switcher and select a slide by number	(Press the slide's number)
Select the next slide in the slide switcher	+ or =
Select the previous slide in the slide switcher	-
Close the slide switcher and show the selected slide	Return
Close the slide switcher without changing slide	Esc
Reset the presenter-display timer to zero	R
Scroll the presenter notes up	U
Scroll the presenter notes down	D
Swap the primary display and secondary display	X
End the presentation	Esc, Q, . (period), or Command+. (period)

Controlling a presentation with the Apple Remote

If you have an Apple Remote, you can use it to control the presentation as follows:

> **Start the presentation.** Activate Keynote, and then press the Play button.

- **Move to the next build or slide.** Press the Next/Fast-forward button.

- **Move to the previous build or slide.** Press the Previous/Rewind button.

- **Display the slide switcher.** Press the Menu button. Press the Next/Fast-forward button or the Previous/Rewind button to select the slide you want, and then press the Play button.

Controlling a presentation from an iPhone or iPod touch with Keynote Remote

When you've set up Keynote Remote to make your iPhone or iPod touch a remote control for Keynote on your Mac, you can quickly run a slide show from the device.

On the Home screen, touch Remote to launch Keynote Remote. Keynote Remote automatically connects to your Mac and displays the Connected to Keynote screen.

Touch Play Slideshow to start the slide show playing. You'll then see the first screen and presenter notes if you're using portrait orientation (see figure 12.9) or the first two slides if you're using landscape orientation (see figure 12.10).

Here's how to run the slide show from the iPhone or iPod touch:

- **Display the next slide.** Swipe your finger from right to left across the screen.

- **Display the previous slide.** Swipe your finger from left to right across the screen.

- **Jump to the first slide.** Touch Options, and then touch First Slide.

- **End the show.** Touch Options, and then touch End Show.

12.9 In portrait orientation, the iPhone or iPod touch can display one slide and any presenter notes attached to it.

12.10 In landscape orientation, the iPhone or iPod touch can display the current slide and the upcoming slide, but no presenter notes for either.

Using other applications during the presentation

If you need to use other applications while giving the presentation, you can simply press H to hide the presentation screens, revealing your desktop and any applications and windows open on it. You can then use the applications as normal — for example, you can launch Safari from the Dock if you need to view a Web site.

When you're ready to return to the presentation, click the Keynote icon in the Dock. Mac OS X drops you straight back into the presentation where you left it.

Genius

If you've selected the Allow Exposé, Dashboard and others to use screen check box in Keynote's Slideshow Preferences, you can also use the keyboard shortcuts, mouse shortcuts, or hot corners for these tools to reveal other applications and access them. For example, use your Exposé All Windows shortcut to display miniatures of all your open windows so that you can click the one you want.

Creating Handouts of Your Presentation

Often you'll want to give your audience handouts of your presentation, either so they can get up to speed on the topic while waiting for you to start or at the end of the presentation as a takeaway to reinforce your message.

Keynote makes creating a handout easy:

1. **Open the Print dialog box.** Choose File ⇨ Print or press ⌘+P.

2. **If the Print dialog box is collapsed to its small size, hiding most of the options, expand it by clicking the disclosure triangle to the right of the printer's name at the top.**

3. **In the Print area, select the Handout option button (see figure 12.11).**

4. **In the Handout pop-up menu, choose the number of slides per page.** The preview on the left shows how the pages will look, and the page readout shows how many printed pages there will be. You can click the arrow buttons to move from page to page in the preview.

5. **Choose options for the handout by selecting and deselecting the check boxes below the Handout pop-up menu.** You can add divider lines between the slides, print at draft (lower) quality to save time and ink or toner, include your presenter notes, and add rule lines.

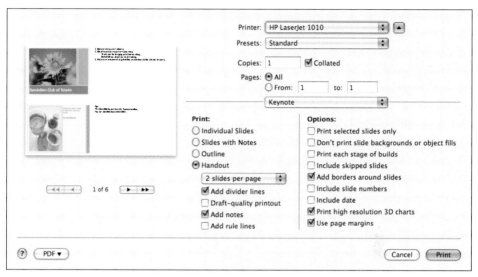

12.11 When printing handouts from your presentation, choose the number of slides per page and decide whether to include presenter notes.

6. **In the Options area, select the check boxes for the options you want to use:**

   - **Print selected slides only.** Select this check box to print just part of your presentation.

   - **Don't print slide backgrounds or object fills.** Select this check box to omit the backgrounds and fills so that the text is easier to read. This setting is helpful for printing colorful presentations in grayscale.

   - **Print each stage of builds.** Select this check box if you want to print a separate slide for each stage of a build. This can be helpful when you're creating a handout for the presenter to use when preparing the presentation, but the audience doesn't usually need to see the builds.

   - **Include skipped slides.** Select this check box to include any slides in the presentation (or in the selection) that are marked for skipping. This setting is useful when you need to provide the audience with more information than the presenter can deliver in the allotted time.

   - **Add borders around slides.** Select this check box to print a thin border around each slide. This is usually a good idea, especially for slides with light (or white) backgrounds you're printing on white paper.

329

- **Include slide numbers.** Select this check box if you want to include the slide numbers for reference (for example, so that you can refer back or forward to a slide by number).

- **Include date.** Select this check box to include the date — useful for reference purposes.

- **Print high resolution 3D charts.** Select this check box to print 3D charts as clearly as possible. If your presentation contains 3D charts, this is usually a good idea.

- **Use page margins.** Select this check box to make Keynote obey the page margins — in other words, *not* use them by putting slides or text in them. Deselect this check box if you want Keynote to use the whole page, but be aware that most printers will cut off text or objects that are placed right at the edges of the page.

7. **Choose the number of copies in the Copies box.**

8. **Click the Print button to print the handout.**

Genius

Sometimes you may find it useful to print just the outline of a presentation rather than the slides. To do so, open the Print dialog box by choosing File ➪ Print or pressing ⌘+P, select the Outline option button in the Print area, and then click Print.

Creating a Presentation That Plays Automatically

Live and in person is usually the most effective way of giving your presentation, because you can adapt the presentation to the audience's needs, gauge their reactions, and take questions to make sure you've nailed every point. But when you can't be there, a presentation that plays automatically is a fair substitute.

To create a presentation that plays automatically, build the presentation as usual using the techniques described in Chapters 10 and 11. You'll probably want to set up most of the builds and transitions to run automatically after a delay you specify, but you can also set a default delay for builds and transitions set to run on a click, as you'll see in a moment. If the presentation needs a sound track, give it one, or record narration.

Then set up the presentation to run automatically like this:

1. **Open the Document pane of the Document Inspector (see figure 12.12).** Click the Inspector button on the toolbar, click the Document Inspector button, and then click the Document tab.

2. **In the Presentation pop-up menu, choose Self-playing.**

3. **In the Delay area, set the delays to use for all transitions and builds set to run on click.** Usually, you'll want a much shorter interval for builds than for transitions.

4. **In the Slideshow Settings area, choose further settings as needed:**

- **Automatically play upon open.** Select this check box if you want Keynote to launch the presentation as soon as someone opens it.

- **Loop slideshow.** Select this check box to make Keynote keep playing the presentation over and over again until you stop it.

- **Restart show if idle for.** If you're letting visitors run the slide show in their own time, select this check box and set a low number of minutes (m) — for example, 3 m or 5 m. If a visitor abandons the slide show before the end, Keynote waits for the timeout, and then starts the show from the beginning.

12.12 Use the Document pane of the Document Inspector to create a self-playing presentation and to protect it with a password if necessary.

- **Require password to exit show.** To prevent a competitor from sabotaging your kiosk's looped presentation, select this check box. When the time comes to close the show, you'll need to enter your username and password for the Mac before Keynote will relinquish its grasp on the screen.

5. **Close the Document Inspector, save the presentation, and then test that the delays are suitable for the transitions and builds.** Adjust the delays if necessary.

Sharing a Presentation in Other Ways

Delivering a presentation from Keynote is great because you have total control over what your audience sees and hears. But Keynote also makes it easy to share your presentation in other ways — everything from creating a PowerPoint slide show to publishing it on YouTube or on iWork.com, to turning it into a podcast that you can distribute via the iTunes Store or other online channels.

Table 12.2 summarizes when to use these different sharing options.

Table 12.2 Ways of Sharing Your Presentations

Sharing Type	Share Your Presentation in This Way When
Share on iWork.com	You want to make the document available for other iWork.com users to comment on or download.
Share via Mail	You want to send the document via email.
Send to iDVD	You want to burn the presentation to a DVD so that you can distribute it on a physical medium.
Send to iPhoto as slides	You want to export slides as image files so that you can view them in iPhoto or share them individually.
Send to iWeb as a PDF file	You want visitors to your iWeb Web site to be able to view the presentation as a static document.
Send to iWeb as a Keynote document	You want visitors to your iWeb Web site to be able to download the presentation and edit it or view it in Keynote.
Send to iWeb as a video podcast	You want visitors to your iWeb Web site to be able to download the presentation as a podcast they can view in iTunes or on an iPod or iPhone.
Send to YouTube	You want to make a movie of the slide show available for download on the popular video-sharing site.
Create a PDF file	You need to share a presentation as a printable document. You can choose between including just the slides or the slides along with their notes.
Create a PowerPoint presentation	You need to work with people who use Microsoft PowerPoint (either on Windows or on the Mac).
Create images	You want to create images of individual slides that you can share as needed (without involving iPhoto).
Create an HTML document	You want to create a presentation that consists of a series of linked Web pages with images.
Send to iTunes	You want to create a movie version of the presentation that will play in iTunes, on the iPod, or on the iPhone.

Saving a presentation as a PowerPoint slide show

When you need to share a Keynote presentation with someone who uses Microsoft PowerPoint, you can export or save the presentation as a PowerPoint file with just a few clicks like this:

1. **Open the presentation and choose Share ⇨ Export to display the Export dialog box.**

2. **Click the PPT button to display the PowerPoint pane.** This pane contains no options.

3. **Click the Next button to display the Save As dialog box.**

4. **Enter the name for the PowerPoint file in the Save As box, choose the folder, and then click Export.**

Genius

At this writing, PowerPoint has no conversion filter for importing Keynote presentations. So if you need to move a presentation from Keynote to PowerPoint, use Keynote's export feature, and then adjust any slides that don't convert perfectly.

This Share ⇨ Export method is easy enough, but Keynote also provides a couple of alternatives:

- **Use the Save As dialog box.** You can also create a PowerPoint version of a presentation by choosing File ⇨ Save As, selecting the Save copy as check box, choosing PowerPoint Presentation in the pop-up menu, and then clicking Save. This has the same result as Share ⇨ Export; it's just a different way of approaching the same task.

- **Create and send a PowerPoint presentation.** If you want to save a Keynote presentation as a PowerPoint file and then immediately send it via Mail, you can choose Share ⇨ Send via Mail ⇨ PowerPoint. Keynote exports the presentation to a PowerPoint file, starts a new email message, and attaches the PowerPoint file to it. This is convenient, but usually it's best to check through the files you export to PowerPoint to make sure everything looks right before you distribute them.

Genius

If you have a PC available but not Microsoft PowerPoint, download the free PowerPoint Viewer from the Microsoft Web site (www.microsoft.com/downloads) to check the Keynote presentations you export to PowerPoint format. The Viewer lets you open and view PowerPoint presentations, but not edit them.

Exporting a presentation to a QuickTime movie

If you want to turn your presentation into a movie file that will play on both Windows and on Macs, create a QuickTime movie like this:

1. **Choose Share ⇨ Export, and then click the QuickTime button.** Keynote displays the Create a QuickTime Movie dialog box (see figure 12.13).

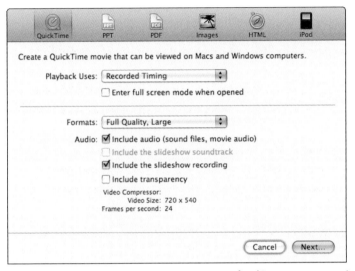

12.13 Creating a QuickTime movie is a great way of making your presentation playable on any Windows PC or any Mac.

2. **In the Playback Uses pop-up menu, choose how playback will run:**

   - **Manual Advance.** The viewer gets to move from slide to slide.

   - **Hyperlinks Only.** The movie includes hyperlinks that the viewer clicks to move from slide to slide.

   - **Recorded Timing.** The movie uses the timing you've recorded into it.

   - **Fixed Timing.** The movie advances on-click slides after the time you enter in the Slide Duration box and on-click builds after the time you enter the Build Duration box. Slides and builds with automatic timing use that timing.

3. **If you want QuickTime to play the movie full screen, select the Enter full screen mode when opened check box.**

4. **In the Formats pop-up menu, choose the format you want to use:**

- **Full Quality, Large.** This size uses 720×540 resolution and 24 frames per second, which is enough for full-screen video to look good. The disadvantage is that files can become large for transferring over the Internet.

- **CD-ROM Movie, Medium.** This size uses 360×270 resolution and 12 frames per second. This size looks fine playing in a window but looks rough and choppy playing full screen.

- **Web Movie, Small.** This size uses 180×135 resolution and 12 frames per second. This size looks okay playing in a small area of a Web page, but you lose a considerable amount of detail.

- **Custom.** Use the Custom QuickTime Settings dialog box to create exactly the video size and frame rate you want. You can also choose which type of video compression to use.

5. **In the Audio area, select the audio items you want to include in the movie.** Depending on the audio you've added to the slide show, you can select among items including sound files and movie audio, the slide show soundtrack, or a slide show recording you've made.

6. **Select the Include transparency check box if you want the movie to include any transparency effects you've used in the slide show.**

7. **Click Next, choose the filename and location in the dialog box that appears, and then click Export.**

Creating a PDF file of a presentation

A PDF file lets you create an online document that contains the presentation in a format that can be viewed on almost any computer or operating system.

Here's how to create a PDF from a presentation:

1. **Choose Share ⇨ Export to open the Export dialog box.**

2. **Click the PDF button to display the PDF pane if any of the other panes is displayed.** Figure 12.14 shows the PDF pane with the Security Options section displayed.

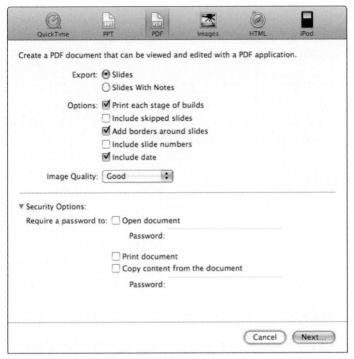

12.14 The PDF pane of the Export dialog box lets you choose which slides and options to include, which image quality to use, and whether to apply security.

3. **In the Export area, choose what to export by selecting the Slides option button or the Slides With Notes option button.**

4. **In the Options area, choose whether to print each stage of builds, include skipped slides, add borders, and include the slide numbers or the date.**

5. **In the Image Quality pop-up menu, choose the image quality you want: Good, Better, or Best.**

   ● Best is normally what you'll want, because it produces a full-quality PDF. Keynote keeps each image at its full resolution.

   ● If you find that Best produces files that are too large for the way you're using to distribute them, experiment with the Better setting or the Good setting to produce a smaller file. Better reduces the image quality to 150 dots per inch (dpi); Good uses 72 dpi.

6. **If you want to secure the PDF file, click the Security Options disclosure triangle to reveal the security options.** You can then require a password to open the document or a (different) password to print it or copy information from it.

7. **Click Next.** Keynote displays a Save dialog box that lets you choose the name to give the PDF and the folder in which to save it.

8. **Click Export.** Keynote displays a progress readout as it exports the PDF.

After Keynote finishes the export, open a Finder window to the folder in which you saved the PDF. Click the file, check its file size, and then double-click the file to open it in Preview (or your default PDF viewer). Make sure that the document appears the way you want it, and that any security options you chose are working effectively, before you distribute the file.

Creating image files from your slides

Keynote lets you create image files of the slides in a presentation two ways:

- ◉ Choose Share ⇨ Send To ⇨ iPhoto.
- ◉ Choose Share ⇨ Export to open the Export dialog box, and then click the Images button at the top.

Either way, you see the controls shown in figure 12.15. Choose settings as follows:

1. **In the Slides area, choose which slides to export. Select the All option button if you want them all.** Otherwise, select the From option button and enter the starting and ending numbers in the boxes — for example, from 6 to 11.

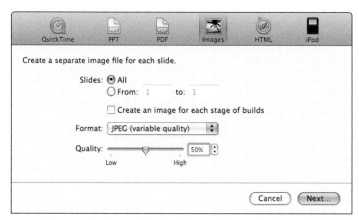

12.15 When creating image files for iPhoto or for another destination, choose which slides to include, whether to include builds, and the format and quality for the images.

2. **Select the Create an image for each stage of builds check box if you want a sepa-rate image for each build.** This is handy if you will use the images to show the presen-tation as the audience would normally see it.

3. **In the Format pop-up menu, choose JPEG, PNG, or TIFF, as appropriate.** If you choose JPEG, drag the Quality slider to specify the quality. The higher the quality, the better the image looks, but the larger the file size is, so you may need to experiment with exporting images at various qualities so that you can determine which quality you need.

Genius

Use JPEG files when you want to produce widely readable files with a smaller file size and you're prepared to lose some quality. Use PNG files when you need high-quality files for computer use. Use TIFF files when you need high-quality files for print publishing.

4. **Click the Next button.** If you're sending the pictures to iPhoto, Keynote prompts you to specify the new album name. If you're saving the pictures to a folder, Keynote prompts you to choose the folder.

Exporting a presentation to Web pages

Sometimes you'll want to turn a presentation into Web pages without sending it to an iWeb Web site. To do so, choose Share ⇨ Export, and then click the HTML button. The options are the same as those for creating image files (discussed in the previous section), but you also get to choose whether to include navigation controls (Home, Previous, and Next links).

Publishing a presentation on YouTube

YouTube is the most popular video-sharing site on the Web, so it's a great place to put any presen-tation you want to share with the whole world.

Here's how to share a presentation on YouTube:

1. **Choose Share ⇨ Send To ⇨ YouTube.** Keynote displays the Send your slideshow as a movie to YouTube dialog box (see figure 12.16).

2. **Type your YouTube account name in the Account box and your password in the Password box.**

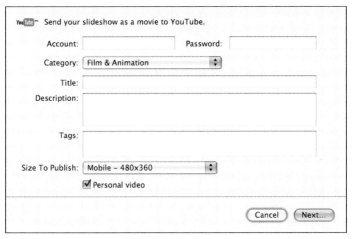

12.16 Keynote lets you turn a slide show into a movie and publish it to YouTube in moments.

3. **Open the Category pop-up menu and choose the YouTube category into which you want to put the movie.** The categories are largely self-explanatory. For example, if your slide show is a demonstration of cooking, you might choose Howto and Style or Education. If it's a travelogue, Travel & Events is usually the best bet.

4. **Type the slide show's title in the Title box.** Make it as snappy, descriptive, and memorable as possible.

5. **Type a description of the slide show in the Description box.** The challenge is to describe the slide show enough to grab viewers while keeping it concise. Put the most important information first to grab the attention of anyone browsing YouTube videos.

6. **In the Tags box, type the tags you want to give the movie.** Tags are the words with which YouTube matches people's searches, so put in a variety of related terms. Separate each tag from the next with a comma.

7. **In the Size To Publish pop-up menu, choose the size you want to publish:**

   - Select Mobile if you want to post the movie in sizes that'll work on mobile phones, on computers, and in browsers on YouTube.

   - Select Medium if you want a larger size that'll work on Apple TV, computers, and YouTube.

8. **Select the Personal video check box if you want to share it only with people on your lists — either the Friends list and the Family list that YouTube automatically gives you, or lists you've set up yourself.**

9. **Click Next, and follow through the steps for publishing the slide show and either notifying your contacts about it or viewing it yourself.**

Publishing a presentation on iWork.com

When you're collaborating with other users of Apple's iWork.com online service, iWork.com is a great way of sharing your presentations. See Chapter 1 for a walk-through of the mechanics of the process. The Advanced options let you share a Keynote presentation in any or all of four formats: Keynote '09 format, Keynote '08 format, PowerPoint presentation format, or PDF.

Turning a slide show into a podcast

If you want to create a podcast from a slide show, you can send the slide show to GarageBand like this:

1. **Choose Share ⇨ Send To ⇨ GarageBand.** Keynote displays the Create a podcast you can edit in GarageBand dialog box (see figure 12.17).

12.17 You can quickly export a slide show to GarageBand for editing into a slick podcast.

2. **In the Playback Uses pop-up menu, choose Recorded Timing or Fixed Timing, as appropriate.** If you choose Fixed Timing, enter the duration for on-click transitions in the Slide Duration box and the duration for on-click builds in the Build Duration box.

3. **In the Audio area, choose which audio to include.** For a fixed-timing slide show, you can select the Include the slideshow soundtrack check box. For a recorded-timing slide show, you can choose between the Include the slideshow soundtrack check box and the Include the slideshow recording check box.

4. **Click Send, choose the filename and location in the dialog box that appears, and then click Export.**

See *iLife '09 Portable Genius* (Wiley, 2009) for instructions on creating podcasts with GarageBand.

Note

Reducing the Size of a Presentation

If your presentation consists of only text, it'll be compact — but as soon as you add high-resolution pictures, audio files, or movie files, the presentation file can quickly grow to a large size.

If you're creating a presentation that you'll give from your Mac, you may not need to worry about the presentation's size. But if you need to transfer the presentation to another computer or distribute it via the Internet, you will probably want to reduce its file size as much as possible.

If there's a particular image that you know is huge, and you've cropped it or masked it heavily, you can reduce it like this. Click the image, and then choose Format ➪ Image ➪ Reduce Media File Size.

If you just want to squeeze the presentation down as much as possible, choose File ➪ Reduce File Size.

Finish creating your presentation before you reduce its size in these ways. Otherwise, you may find Keynote has clipped off the extra part of the movie (because you had shortened it in the presentation) that you now realize you need to show. If you need to enlarge an image you've made smaller, you can do so, but you lose some image quality.

Glossary

~ (tilde character). Represents your home folder in the Mac OS X file system. See also *home folder*.

absolute reference In Numbers, a reference that always refers to the same cell, even if you move the formula to a different formula cell. An absolute reference uses a dollar sign ($) before the column name and the row number — for example, $A$1 for cell A1. See also *relative reference* and *mixed reference*.

auto-completion A feature that monitors the entries in table columns and suggests an existing entry when you type the first few letters of its name.

Auto-Correction A feature for automatically entering the full version of a term in place of an incorrect or abbreviated version that you type. For example, if you type "teh", the Auto-Correction feature automatically corrects it to "the". You can create Auto-Correction entries to correct your own errors or to let you type long strings of text quickly. Each iWork application has separate Auto-Correction entries, so you can use the same term to enter a different item in each application if necessary.

autofilling In Numbers, automatically entering values in cells based on cells that already contain values. For example, if cell A1 contains Monday, and cell B1 contains Tuesday, you can select those cells and then click and drag to the right to fill in Wednesday in cell C1, Thursday in cell D1, and so on.

bookmark In Pages, an invisible marker that you can place to identify a particular point or section of a document. You can then move quickly to the bookmark or link another part of the document to it.

category In Numbers, a way of organizing rows within tables. A category is like a heading within the table, and you can collapse the category to hide the rows it contains.

cell format A set of formatting that tells the iWork application what kind of contents a table cell has. For example, you can display a currency with the U.S. dollar symbol ($), two decimal points, and the thousands separator, making the cell display 99999.99 in the easier-to-read $99,999.99 format.

Comments pane In Pages, a pane that appears at the left side of the window and shows the comments and the tracked changes in the document. Pages automatically displays the Comments pane when you turn on Track Changes and when you insert a comment.

conditional formatting A type of formatting rule you apply to a table cell to make the iWork application monitor it for unusual values. When such a value appears, the application formats it the way you have chosen. For example, you can make suspiciously low values appear in red so that you can pick them out immediately in a spreadsheet.

CSV The abbreviation for Comma-Separated Values, a way of storing spreadsheet information in a text file. The contents of each cell are delimited (separated) with commas.

direct formatting In Pages, formatting that you apply to text without using a style. For example, you can apply boldface or italic to a word. See also *style*.

document preview A small PDF saved in the iWork document that lets Mac OS X's Quick Look feature display the document accurately. The preview enables you to recognize your documents more easily than the Quick Look feature's default view of a document does.

filter In a table, to narrow down the rows displayed so that you see only the ones that match your criteria.

floating object In Pages, an object such as an image that you can place freely on the document's page rather than anchoring it in the text stream. See also *inline object*.

footer In Pages or Numbers, text that appears at the bottom of each page — for example, the page number and total number of pages (such as Page 2 of 14) or the author's name. See also *header*.

Format bar A bar that appears across the top of each iWork application's window and provides widely used formatting commands for the selected object. As you select different objects, some of the commands on the Format bar change.

formula A custom mathematical calculation that you build out of values, references, and formulas to make a table cell perform exactly the calculation you need. See also *function*.

Formula bar In Numbers, a bar that appears below the Format bar and enables you to quickly create and edit formulas.

Formula Editor In Numbers, a window that you use to edit complex formulas.

Formula list In Numbers, a pane that you can open at the bottom of the Numbers window to review a list of the formulas in the spreadsheet.

function A predefined mathematical calculation that you can apply to a table cell. For example, the SUM() function adds all the values in the cells you choose. See also *formula*.

header In Pages or Numbers, text that appears at the top of each page — for example, the page number, document name, and date. See also *footer*.

home folder The folder in your Mac's file system that Mac OS X assigns to your user account. Your home folder contains the files for your user account, including the Documents folder, Music folder, and Movies folder. Mac OS X uses the ~ (tilde) character to represent your home folder. For example, ~/Documents refers to the Documents folder in your home folder.

hyperlink A text item or other object in a document that you can click to jump to a linked location. Hyperlinks often lead to Web sites, but the iWork applications also let you create hyperlinks to other documents and hyperlinks that automatically start creating a new email message.

inline object In Pages, an object such as an image that you insert inside the text stream of a document. Pages then treats it as a character in the text, so if you add text before the inline object, it moves toward the end of the document. See also *floating object*.

Instant calculation results In Numbers, an area at the lower-left corner of the window that automatically calculates several widely used calculations (sum, average, minimum, maximum, and count) when you select two or more cells in a table. You can drag the formula for one of the calculations to a cell in which you want to use it.

invisibles In Pages, formatting marks — such as spaces, tabs, and paragraph breaks — that normally are invisible on-screen, although you see their effects. You can display invisibles by opening the View pop-up menu on the toolbar and choosing Show Invisibles.

iWork.com Apple's online service for collaborating on documents created in the iWork applications. Click the iWork.com button on any iWork application's toolbar to connect to iWork.com.

jellybean button The jellybean-shaped button at the right end of the title bar in many Mac application windows. You can click this button to hide the toolbar or display it again.

layout view In Pages, a view that shows the outlines of a document's text areas, enabling you to easily see the headers, footers, text boxes, and other elements. See also *writing view*.

light table view In Keynote, a view that shows a thumbnail version of each slide, enabling you to get an overview of the presentation or rearrange the slides.

mail merge The process of inserting standardized records, such as names and addresses from Address Book, in merge fields in a Pages document to create a separate document for each record. For example, you can use mail merge to create a letter to each customer by name.

master slide A layout of text and objects for a slide. The master slide is part of the theme. See also *theme*.

mixed reference A reference that is absolute for either the column or the row, but not for both. See also *absolute reference* and *relative reference*.

navigator view In Keynote, a view that displays the Slides pane on the left of the window with a thumbnail of each slide.

Normal view In Numbers, a view that displays the current sheet without headers, footers, or page breaks, letting you concentrate on the sheet's contents rather than its layout. See also *Print view*.

OCR See *optical character recognition*.

operator A symbol you use to tell Numbers which operation to perform on the value or values in a formula. For example, in the formula =B3/B2, the forward slash (/) is the division operator and tells Numbers to divide the value in cell B3 by the value in cell B2. See also *formula*.

optical character recognition (OCR) A technology for automatically identifying the characters in an image, enabling you to turn a scanned document into text.

outline view In Keynote, a view that replaces the Slides pane on the left of the window with an outline of the text on each slide, enabling you to quickly edit the presentation's text.

page layout document In Pages, a document — such as a newsletter, brochure, poster, or business card — that has text in various discrete sections, mostly in separate text boxes. Many page layout documents also include many graphical objects, such as images and tables. See also *word processing document*.

placeholder A predefined area into which you can insert an object. For example, many page layout documents in Pages contain placeholders for images.

plain text A file format that contains only text with no formatting or objects (such as images). Plain text files are small, and almost every text editor and word processor can open them successfully.

presenter notes In Keynote, notes that appear for the presenter of the presentation, not for the audience.

Print view In Numbers, a view that shows how the spreadsheet will look when divided into pages for printing. See also *Normal view*.

Proofreader In Pages, a feature for automatically checking the grammar and style of a document.

Quick Look The Mac OS X feature that lets you quickly preview the contents of a document from a Finder window without opening the document. To use Quick Look, select the document and press spacebar, or click the document and click the Quick Look button on the toolbar.

relative reference A reference that is relative to the formula cell's position in the table. If you enter the formula =B2+B3 in cell B4, it adds the values in cell B2 and cell B3; if you move the formula to cell C4, Numbers changes it to =C2+C3. See also *absolute reference* and *mixed reference*.

Remote application An application for the iPhone and iPod touch that enables you to control a Keynote presentation from one of these devices.

RTF The abbreviation for Rich Text Format, a file format used for exchanging text-based documents that contain formatting. See also *plain text*.

Search sidebar In Pages, a panel that you can use to search for all instances of a word or phrase in your document, and view a list of the results. You can click a result in the list to view it on the page.

section A part of a Pages document. You can quickly insert a new section from the Sections pop-up menu on the toolbar.

sheet In Numbers, a worksheet on which you can place tables and other objects as needed. See also *table*.

sheet canvas In Numbers, the area of the window on which you manipulate a sheet.

Sheets pane In Numbers, an area of the window that you use to select or manipulate sheets, tables, or charts.

slide navigator In Keynote, a pane that displays thumbnails of all the presentation's slides. You can click the slide you want to display.

slide only view In Keynote, a view that hides the Slides pane or Outline pane to provide more room for viewing and editing a single slide.

sort In a table, to rearrange the rows into a different order by the contents of one or more columns. You can choose between an ascending sort (from A to Z, from smallest to largest, and from earliest date to latest) and a descending sort (the opposite).

Spotlight Mac OS X's search technology that you can access by clicking the magnifying-glass icon at the right end of the menu bar. You can make iWork documents easier to find with Spotlight by adding metadata such as an author name, a title, keywords, or comments.

style A collection of formatting that you can apply quickly to a paragraph, selected text, list item, or table of contents (TOC) item.

style override In Pages, applying direct formatting (such as boldface or strikethrough) to text, changing the formatting of the style applied to the text. See also *direct formatting*.

Styles drawer In Pages, a panel that slides out of the window to display a list of styles. The Styles drawer can appear on either the left side or right side of the Pages window, depending on where space is available on-screen to display it.

Styles pane In Numbers, a pane on the left side of the window that you use to apply predefined formatting to cells or tables.

table In Numbers, a range of cells on a sheet that you treat as a single unit. A sheet can contain one or more tables.

table style In Numbers, a complete set of formatting for a table, including everything from the text formatting for body cells to the border formatting for header and footer cells, plus any background image or color the table as a whole needs.

template A file that contains the prebuilt skeleton of a Pages document or Numbers spreadsheet. A template can include anything from a minimal amount of data or formatting to an almost finished document.

text box A container that you can use to place text exactly where you need it to appear in a document. You can link two or more text boxes in a chain and flow text through them.

theme A file that contains the look and feel of a Keynote presentation. Each theme contains the master slides for various types of slides, from title slides to slides with images and bullet points.

Thumbnail view In Pages, a sidebar that appears at the left side of the window showing a thumbnail picture of each page. You can click a thumbnail to go to its page.

Time Machine Mac OS X's automatic backup feature for creating copies of essential files on an external hard drive or Time Capsule network-backup drive.

Track Changes In Pages, the feature for applying marks to changes such as insertions, deletions, and formatting changes so that you can easily tell what has been altered.

transition In Keynote, a visual effect that occurs during the switchover from one slide to another — for example, a rotating-cube effect or a cross-dissolve effect.

word processing document In Pages, a document — such as a report, letter, or resume — in which most of the text is in one main section that flows from page to page as needed rather than being laid out in separate text boxes. See also *page layout document*.

writing view In Pages, a view that shows the document's contents but not the outlines of the text areas. See also *layout view*.

Index

The Genius is in.

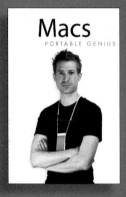

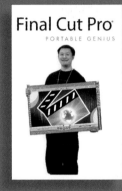